Im/partial
Science

RACE, GENDER, AND SCIENCE
Anne Fausto-Sterling, General Editor

Feminism and Science (1989)
Nancy Tuana, Editor

The "Racial" Economy of Science: Toward a Democratic Future (1993)
Sandra Harding, Editor

Im/partial Science: Gender Ideology in Molecular Biology (1995)
Bonnie Spanier

*The Less Noble Sex: Scientific, Religious, and Philosophical
Conceptions of Woman's Nature*
Nancy Tuana

Love, Power and Knowledge: Toward a Feminist Transformation of the Sciences
Hilary Rose

Women's Health—Missing from U.S. Medicine
Sue V. Rosser

Deviant Bodies: Critical Perspectives on Difference in Science and Popular Culture
Jennifer Terry and Jacqueline Urla, Editors

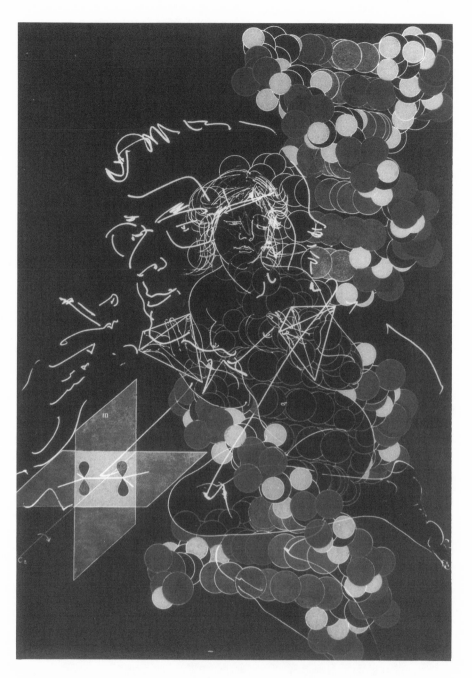

"Portrait of Professor Woodward," lithograph by Hans Erni, 1977, commemorating the sixtieth birthday of Robert Burns Woodward, Harvard University Chemistry Department. Reproduced by permission of the artist.

Im/partial Science

Gender Ideology in Molecular Biology

Bonnie B. Spanier

Indiana University Press
Bloomington and Indianapolis

Frontispiece: lithograph, "Portrait of Professor Woodward," by Hans Erni, commemorated the sixtieth birthday of Robert Burns Woodward of the Chemistry Department of Harvard University. It was displayed in Harvard's undergraduate science center in 1978. Reproduced with permission of the artist and the assistance of the scientist's daughter, Crystal Woodward.

The paper used in this publication meets the minimum requirements of American National Standard for Information Sciences—Permanence of Paper for Printed Library Materials, ANSI Z39.48-1984. ⊚

Manufactured in the United States of America

Library of Congress Cataloging-in-Publication Data

Spanier, Bonnie.
Im/partial science : gender ideology in molecular biology
/ Bonnie B. Spanier.
p.   cm. — (Race, gender, and science)
Includes bibliographical references (p.   ) and index.
ISBN 0-253-32892-6 (cl). — ISBN 0-253-20968-4 (pbk.)
1. Molecular biology.   2. Sexism in biology.   3. Feminism.
I. Title.   II. Series.
QH506.S66   1995
574.8'8'01—dc20   94-48044
1 2 3 4 5   00 99 98 97 96 95

For
Jane M. Oppenheimer,
extraordinary scientist and teacher,
and for all similarly inspiring teachers

Contents

PREFACE

How I Came to This Study

Scanning the science section in the University at Albany's bookstore a few years back, I pulled from the shelf an attractive new molecular biology textbook entitled *Molecular Cell Biology*. The three authors, James Darnell, Harvey Lodish, and David Baltimore, were familiar to me as leading figures in molecular biology and virology, my fields of graduate study. In fact, as a very green and eager graduate student at Harvard, I had the pleasure and challenge of working in the MIT lab of David Baltimore, who later received a Nobel Prize. Opening the six-pound textbook, costing $42.95, I was, in a way, coming home. But I had changed. I had removed old lenses, so that I saw what had been invisible to me and to many others: the distorting effects of cultural biases about gender.[1]

It was almost a decade since I had left laboratory research on the molecular biology of viral infections, had left the equally rewarding teaching of biology, biochemistry, molecular and cell biology, microbiology, and immunology to pursue three simple goals: increasing the public's access to science; answering the simple question of why a world-famous embryologist, my mentor, had spent her career at a small women's college; and following a complex desire to expand my skills, creative scope, and knowledge beyond the science to which I was trained and from which I had received ample rewards. A privileged daughter, I had a faculty appointment at the State University of New York at Albany, where I was the Women's Studies Program Director, equivalent to a department chair, but in a relatively young interdisciplinary field.[2] At the moment that I discovered the textbook, I was starting a year's leave, supported by an affirmative action program for untenured faculty, to advance my research on the impact of feminism in the natural sciences. My gender consciousness governed the perspective of my engagement with that textbook and, consequently, dramatically changed the focus of my research.

In that moment in the summer of 1987 and in the years following, I brought together three major factors that shape my view of the world: an education in classical biology and biochemistry, credentials and experience in molecular biology, and a feminist awareness of the significance of gender in society and the importance of placing knowledge into its historical and political contexts. This effort at synthesis, this lifting of the constructed barriers separating these three aspects of my being, has been an experience that I hope proves useful to others as well.

I fell in love with DNA and enzymes sometime before junior high school, when I assembled a layered representation of a cell for a science project, researching and painting the different organelles and subcomponents of an animal cell on sheets of clear plastic. Since then, with the encouragement and support of family, friends, husband, and lovers, I have long enjoyed thinking and learning about the world at the level of cells and organelles and molecules, especially the giant biological molecules, macromolecules (such as enzymes, DNA, phospholipids, RNA, and proteins other than enzymes) that are characteristic of life on this planet.

When I was an avid biology and chemistry student in high school in a suburb of New York City, microbiologist Mary (Polly) Bunting was on the cover of *Time* or *Newsweek* as the new president of Radcliffe College. I remember being instantly impressed. She was a scientist, she had built her own home with her husband, and she was raising their children by herself after his death. This role model was all I needed, it seems in retrospect, for me to ignore the stereotyping of woman and scientist as contradictory terms. (Indeed, I later learned of Dr. Bunting's pronouncement: "I decided many years ago that I was far more interested in being a fact than in living anyone else's theory.")[3]

With the help of a National Merit Scholarship, I attended an elite, traditionally white women's college with strong science departments and a commitment to liberal arts education and the love of learning for its own sake (perhaps more typical of college experience in the '60s than can be imagined by students today). At Bryn Mawr College I found another inspiration, the brilliant embryologist and teacher Jane Oppenheimer, who mentored me in a number of ways, including teaching me both research skills and care and respect for animals used in experiments. She applied her very high expectations of students to herself, always pushing beyond current boundaries by combining perspectives from classical biology with insights from new experimental techniques and approaches to the study of life. She also made us aware of one implicit hierarchy: one of her specialties was to dethrone mammals, primates, and humans as the pinnacle of evolution—in favor of birds.

Miss Oppenheimer (we called all the female professors of her generation "Miss" and all the younger, male professors "Doctor," even though all had doctorates) engaged in her history of science research "on the side," as she said, always claiming it was an avocation for her, while her colleagues recognized those writings as significant contributions to the history of biology. However, as she did not teach history of science courses at Bryn Mawr when I was there, I gained only a limited perspective on historical studies of science. During the period of my college years (1963–67), molecular biology was the rage, but the requirements for the biology major at Bryn Mawr continued to emphasize a classical approach to biology: intensive study of invertebrate as well as vertebrate physiology, classification of organisms, embryology, and traditional biochemistry. As a consequence, I developed an appreciation for the diversity of living beings as well as of macromolecules.

Although I was a white, middle-class, female graduate of a prestigious women's college, my access to graduate school in the late 1960s in biochemistry and, eventually, the molecular biology of viruses was made possible as much by the post-Sputnik push for the U.S. to gain supremacy in science and technology as it was by the efforts for equal opportunity for women, which became visible in the second wave of the women's liberation movement beginning in the late 1960s.[4] I eagerly sought research jobs as an undergraduate and was mentored first by Jane Oppenheimer and later by biochemist W. Eugene Knox in Harvard University's Department of Biological Chemistry located at Harvard Medical School.

Eventually, in Michael A. Bratt's lab in Harvard's Department of Microbiology and Molecular Genetics, I found the perfect complement to elegant and abstract molecular biology with my growing concern for improving the quality of life for others. We attempted to understand the molecular workings of disease by studying interactions of viruses (specifically, different strains of the paramyxovirus Newcastle disease virus) with the cells they infect. Having persevered with the help of several mentors, I learned quickly what it means to be an insider in science, being asked to represent our lab at major animal virology conferences and then gaining access to opportunities for postdoctoral fellowships at prestigious research centers.

By 1975, doctorate in hand, I had a position in the Biology Department at Wheaton College in Massachusetts, a small women's college which felt comfortably like Bryn Mawr. As most young researcher/educators do, I applied for research grants. With the support of my thesis advisor and a gifted colleague and friend, Chuck Madansky, and after the usual first rejection, I was fortunate to obtain two coveted grants for young investigators, one from the National Institutes of Health and the other from the American Lung Association.

Just before this time, something had tugged at my sleeve. In my search for grants I had come across a new program at the Bunting Institute at Radcliffe College, renamed around that time for the same Dr. Polly Bunting who had inspired me; she had started the Radcliffe Institute in 1960 to allow women with children some free time and space to pursue their creative or educational dreams. The Bunting grant offered two years of postdoctoral-level support and research expenses for studies of women in American society. Never actually expecting to get the grant, I proposed to interview three women scientists about their work and to write about their science in their own words, making women scientists visible and making science interesting to a wider public. I also planned to take courses in the history of science, a specialty at Harvard.

I was motivated by my respect for and curiosity about Jane Oppenheimer. Knowing that she had received her doctorate from Yale and had spent sabbaticals at Johns Hopkins, I wondered why she had stayed at Bryn Mawr, where the teaching load was much greater and the resources and prestige much less than that at research universities. The question suggests the degree of my ignorance about women's education and opportunities, particularly in the sciences. Such ignorance was not unusual, however, even for a graduate of a (white) women's college, since the women's colleges I knew were not hotbeds of feminist consciousness-raising in the early years of the women's movement. And it was not until 1982 that Margaret Rossiter published her eye-opening historical study of women in American science. Although important feminist scholarship had been published by 1977, I was not at all familiar with it, having spent the previous decade being trained as a scientist at some of the best institutions of higher learning in the United States.[5]

Unexpectedly, Radcliffe offered me the two-year grant at the same time that the National Institutes of Health and American Lung Association grants came through. The choice was difficult because I would be forced as an untenured faculty member to give up my tenure-track position.[6] I chose the Bunting grant, opting for time to develop a different part of me, to learn to see science differently, and to learn about my foremothers. At the Bunting Institute, I met women scholars and artists, nearly forty of them, deeply committed to their work. Among them were activist feminists from whom I began to learn the origins of feminism, including the kind I had absorbed—primarily an individualistic, liberal feminism. From those women I learned about the history of white women, of African American and Native American women and men, of individual and group struggles for the privileges I took for granted, and of the struggles against discrimination and abuse that I had only mildly experienced. Thus, I became conscious of the politics of gender, race, and class, most specifically sexual politics, the term Kate Millett coined to denote power relations ingrained in contemporary social relations and based on assigned gender and the cultural meanings of male and female.[7] Donna Haraway's more recent elaboration captures the extreme nature of sexual politics as "a polyvalent term covering a host of life-and-death issues and struggles for meanings."[8]

While I was working in quaint Cambridge, Massachusetts, a white, middle-class woman was raped and murdered in broad daylight in a park near the Bunting Institute. One of the gifted writers at the Institute flew across the country to testify at the trial of the man who had raped her and had walked her into the hills near the writers' colony to kill her, as he had done to other women. She had managed to escape, had lived to tell her story and help convict the rapist-murderer. Several black women, each walking late at night in Boston's Back Bay area where I was staying, were murdered and the killers never found. I learned nearly firsthand that the struggles around gender, race, and class for meanings in society have deadly consequences for many of us.

Meanwhile, I was auditing courses at Harvard in the history and social study of science, especially Everett Mendelsohn's legendary classes. I found Ruth Hubbard's Biology and Women's Issues, a course that had grown from a seminar's handful of students to the 100 students who filled the biology lecture hall.[9] Along with women's concerns in science, I discovered women's history, and with that, the absences, the gaping lacunae in my education. Thus I began an education that continues to this day, now within the legitimized and legitimizing academic territory of women's studies, an education that has moved me beyond the solipsisms of privilege to recognize the interconnections among the many struggles for freedom.[10]

It was with that background of classical biology and biochemistry, molecular biology, history and social study of science, and feminism that I opened *Molecular Cell Biology* and read:

> *E. coli* F (male) plasmid determines the difference between male and female strains, and encodes the proteins required for cell-to-cell contact during sexual mating. These plasmids are of great use to experimental molecular biologists— they are essential tools of *recombinant DNA* technology. . . .[11]

Male and female bacteria? The male signifier as an essential tool of research? Here was the tip of the iceberg of gender ideology in molecular biology.

What This Book Is About

In what ways might molecular biology, which includes as subject matter cells, macromolecules, and genes, be affected by traditional beliefs about gender?

This work on gender and ideology in molecular biology examines the reproduction of knowledge about molecular biology. Specifically, I explore what the formal discourse of scientists and scientist-educators in molecular biology reveals about the impact of traditional gender beliefs on this ostensibly nongendered subject matter. To understand science in the context of the social, economic, and political matrix with which it is inextricably enmeshed, I ask: How do scientists present this particular field of science to other scientists and to students? And I ask in what ways this science is shaped by one influential aspect of those societal factors: cultural beliefs about gender and the explicit and implicit values stemming from them.

My preliminary work with molecular biology textbooks indicated that inappropriately gendered language and concepts (such as calling bacteria male and female and using models of dominant-subordinate relationships) are indeed found in introductory textbooks, including those that have been recently revised. My investigation of the ways in which societal values are presented in both textbooks and scientific journals, such as *Science* and *Cell*, gave me a strong impetus to develop a feminist analysis of the field with regard to the presentation and treatment of the interplay of science and society. As I worked through the question of *whether* gender beliefs have left their imprint on the ostensibly nongendered fields of cell and molecular biology, that query changed into *the ways* that such beliefs have affected these fields and *the degree of significance* for biological science and for society.

Organization and Summary

The chapters divide into three parts. The first part provides background information and the methodology of my study. The second part analyzes the formal discourse of molecular biology with regard to the language, paradigms, and major principles of the field, using an awareness of culturally gendered concepts as a way of illuminating biases, distortions, and hidden values in the science of molecular biology. The third part examines formal scientific communications for the treatment of the relationship to society of molecular biology in particular and science in general. *Some readers may want to start with the second part (chapter 4) and simply refer to the foundations set in preceding chapters.*

Chapter 2 places this study into the context of general critiques of gender beliefs, feminist perspectives on the natural sciences and molecular and cell biology in particular, and radical critiques concerning ideologies other than gender that have been identified as problematic for science and for society. This chapter also maps the geography of the biological sciences and provides definitions of molecular biology and related fields. A brief history of molecular biology highlights the shift of focus within the subject matter of molecular biology and sets a context for the significance of the field for biology and for society. As this background information is introductory, portions will be more or less useful for different readers. *However, the section on "Feminist Critiques of Cell and Molecular Biology" provides essential information for my analysis in Part II.*

Chapter 2 sets both scientific and feminist contexts for the questions addressed in chapters 4, 5, 6, and 7: Have scientists modified the guiding principles of molecular biology, set in the 1950s, 1960s, and 1970s, in the currently prevailing view of biology? Challenged as inaccurately reductionist, biodeterminist, and hereditarian by radical science critics, has molecular biology changed in the recent decades of accumulated information about differences in gene structure and function among an array of organisms? How does the call by eminent scientists for a "reformulation of a body of related information formerly classified under the separate headings of genetics, biochemistry, and cell biology" affect the predominant approach to the study of life and, more broadly, scientific epistemology?

Chapter 3 describes the methodology and sources I use for the analysis of the current state of molecular biology with regard to masculinist ideology. I discuss the limits and problematics of my study. This chapter also includes a series of "ingredients" or approaches to guide a feminist analysis of any field of science. Scientists and others concerned about extending similar analyses to other areas of science may be particularly interested in the specifics of those methodological issues.

In Part II, chapters 4, 5, 6, and 7 investigate gender ideology and major concepts in molecular biology, identifying and deconstructing the assumptions embedded in the science. Chapter 4 addresses the use of gendered language and gender-associated concepts in the description and approach to the subject matter of the field, particularly with regard to individual cells such as bacteria and eukaryotic reproductive cells. This investigation leads to a feminist analysis of several prevailing paradigms guiding the field.

Chapter 5 looks at the impact of sexual ideology on our thinking about molecules, extending as well to feminized and masculinized brains. Chapter 6 examines an apparent prior commitment to themes as common to Western physiology as they are to molecular biology: single purpose, centralized control, and natural hierarchies. I argue that the language giving primacy to the gene as the controlling element of life is not used only metaphorically; rather, it describes the fundamental principle of the field.

Chapter 7 focuses on organizing or unifying principles in molecular biology. Contrary to the legacy from biology and the other sciences, the "new biology" claims a reformulation of biological knowledge, not around an overarching theory, but around a set of techniques: recombinant DNA technologies. In this chapter, I address the following: How does casting the reformulation of biological knowledge around techniques rather than explicit concepts affect the range of our vision of molecular life? Using the predominant view of cancer from within the field of molecular biology, I explore the consequences for science and society of a powerful technology defining the study of life.

Part III foregrounds the social, economic, and political contexts of molecular biology. Chapter 8 addresses what the current field of molecular biology depicts of the relationship of science to social, economic, and political influence. Here, what is left out is as important as what is presented to readers of journals and textbooks. What are students learning about the role of science in society and the norms of scientific responsibility?

What can be changed in molecular biology, and how might this affect who will do science? Do feminists and nonfeminists interested in science have common concerns with which coalitions can be built? This work calls for a cooperative, even synergistic, relationship between science and feminism.

ACKNOWLEDGMENTS

To say that no achievement represents the work of just one person is not false modesty in my case. This book owes its existence, directly and indirectly, to many people. I am privileged to have had the support of my immediate family, including my sister Leslie Spanier Josephs and my brother Gerard Spanier. Both my parents loved and respected science and believed that I could and should be all I dreamed of being. My loving gratitude goes to them, Irving and Roselyn Solomon Spanier, for their constant support of my development as both scientist and feminist and for their private "grants" which helped me redirect my career.

I would not have succeeded in science without the support, guidance, and inspiration of my science mentors, including Michael A. Bratt, John T. Edsall, Olga Greengard, W. Eugene Knox, Chuck Madansky, and Jane M. Oppenheimer.

I want to thank the Bunting Institute at Radcliffe College and Laura Bornholdt of the Lilly Endowment for the opportunity given to me by the Women in American Society grant program during 1978–80; that fellowship changed my life. My friends and teachers there, especially MaryAnne Amacher, Joyce Antler, Ann J. Lane, Temma Nason, Linda Perkins, Margaret Rossiter, Sue Standing, and Inès Talamantez, enlightened me in ways that still affect me today. Ruth Schmidt brought me back to Wheaton College for one of my most exciting and challenging experiences, the FIPSE-funded Balanced Curriculum Project to incorporate the new scholarship about women into the curriculum. My feminist colleagues at Wheaton, particularly Kathleen Adams, Itala Rutter, Trudy Villars, and the rest of the Feminist Theory Group, supported and stimulated my feminist education. I wish also to acknowledge and thank everyone in the Great Lakes Colleges Association Summer Institute in 1981 for turning my world upside down and inside out. And thank you to the feminist therapists of the world who help us not just survive, but thrive, to enjoy another day and to make the world better for more people.

The Women's Studies Department at the University at Albany has been a model of support and inspiration to me; past and present deans Francine Frank and Judy Gillespie played significant roles as well. I want to express my great appreciation to the University and the State University of New York system for valuing my work and providing grants that have allowed me to pursue this and related work on feminism and science. The NYS United University Professions' Nuala Drescher Award, SUNYA Faculty Research Awards, and the NYS/UUP New and Experienced Faculty Research Awards gave me the time and mental space I needed to complete the book. Members of the

Feminist Theory Discussion Group of the University at Albany Women's Studies Department, especially Joan Schulz, Mary Galvin, Claudia Murphy, Linda Nicholson, and Susan Shafarzek, have made all the difference in the world to my thinking and writing.

Starting with Sandra Harding, Ruth Hubbard, and Rita Arditti, my teachers and colleagues in the growing arena of feminism and science continue to provoke and sustain my development as a scholar in this field. I have gained tremendously from the successful meetings of feminist scientists in such groups as the New England Feminists in Science, the Mellon Seminar on Science at the Wellesley College Center for Research on Women, 1984–85, the Feminists in Science and Technology Task Force of the National Women's Studies Association, and the Albany Area Chapter of the Association for Women in Science. In particular, I wish to express my thanks to Patricia Brown for sharing her knowledge and wisdom with me.

My work has benefited greatly from the constructive suggestions of anonymous and known reviewers and editors. Jill Hanifan gets my very special appreciation for unknotting sentences, reordering jumbled thoughts, and encouraging me to tell my story. I also thank Joan Catapano and Anne Fausto-Sterling for their patience and guidance.

Among the many people who have encouraged me throughout, Ed Herson deserves much credit for insisting that I put aside the constraints of the moment and allow myself to envision a future of my choosing.

And, finally, I want to thank my partner, Joan E. Schulz, for her incisive, if painful, critique of my writing and her ongoing support of my work.

PART I

Context and Methodology

Scientists present biology, the science of life, as impartial and objective. This book offers evidence to the contrary, showing that what counts as mainstream biology—in this case, the biology of life at the cell and molecular levels—is actually a partial vision skewed by invisible biases.

"Gender ideology" refers to a set of predominating beliefs specific to this moment in Western culture, in which male and female are considered a fundamental complementary pair of polar opposites. In this framework, male and female are inherently different from each other, with maleness assumed to be superior and associated with the natural controller, the action initiator, the "brains," as compared to the female as weaker, more passive, inferior. Selective, nonrandom examples from the biology of nonhuman animals have been used to buttress assumptions about "human nature," about the meanings of "male" and "female."

A feminist perspective points to actual males and females, human and nonhuman animals, whose actions contradict predominant stereotypes. In this view, the ideological bent of male domination and gender difference can be critiqued for its unsupported assumptions and the ways in which dominant interests are served as a result. Gender ideology is understood as far more than a reflection of societal views: as a powerful force supporting and justifying material privilege for some and undeserved deprivation of spirit and substance for others.

With so many examples of the influence of biased gender ideology on the construction of knowledge—whether in the form of canons of literature and the arts or social and economic theories that ignore gender, race, and class as organizing principles of society—I ask: How have beliefs about what is natural and what is sex pervaded the content of molecular biology?

Part I lays out the premises of my argument; it also gives some background about molecular biology itself. Part II provides a variety of specific examples of genderized distortions of life at the level of molecules and cells. These examples range from superimposing inappropriate sex onto hormones to privileging one form of relations in cells and, even, to defining "life" with a male-associated twist. Part III focuses on how scientists offer up molecular biology as another disinterested, objective science.

The conclusion comes full circle, starting from the evidence that scientists generally preserve genderized biases in the substance of their science so that the science reinforces distorted beliefs about gender. Who will become a scientist, then, is heavily influenced by what science is done and what values are buried in it. Breaking this detrimental cycle requires changing the values embedded in molecular biology.

1

Molecular Biology from a Feminist Perspective

AN INTRODUCTION

[Molecular biology] is the dependable way to seek a
solution of the cancer and polio problems, the problem
of rheumatism and of the heart. This is the knowledge
on which we must base our solution of the population
and food problems. This is the understanding of life.
—Warren Weaver, 1949[1]

[W]e have complete confidence that further research,
of the intensity recently given to genetics, will eventually provide man with the ability to describe with completeness the essential features that constitute life.
—James D. Watson, 1990[2]

I do not underrate the charge of presumption which
must attend to any woman who will attempt to contro-
vert the great masters of science and scientific inference. But there is no alternative!
—Antoinette Brown Blackwell, 1875[3]

Rationale: The Promise and Challenge of Feminist Analyses

During the past few decades, women's studies scholars have amassed convincing evidence that cultural beliefs about gender, race, and class have strongly influenced our current structures of knowledge. Feminist historians, for example, have proposed that the criteria for establishing historical periods have been based mainly on wars and political elections, events controlled by government and military leaders who are, nearly exclusively, privileged males. Organizing history by these arbitrary periods, it is argued, is biased against women's experiences and influence, resulting in distortions of women's activities and interests. Scholars found that freedom and opportunity for those privileged European women who had gained some measure of self-determination actually declined during the Renaissance. Hence the question: Did women have a Renaissance? And an answer: Not during the Renaissance.[4] Similarly, periods of scientific growth have not necessarily advanced the status of women in science.[5] These confluences are not accidents of history. Rather, we find that those conditions and values that advanced a select group of white men (and were, in turn, advanced by them) acted at the same time to exclude women from the scientific professions—or from equitable access to legal, social, political, and economic resources.

Thus, feminist scholarship has revealed significant problems in the disciplines of the Academy with regard to methodology, accuracy, and theory. As such, that scholarship reveals not only *who* and *what* have been left out of what counts as significant history, literature, politics, etc., but also the ways in which the very process of constructing and codifying knowledge has come to present a partial and distorted view of humanity and the world, rather than the impartial, objective perspective claimed by scholars. Feminists are not the only ones who have revealed systematic distortions of bodies of knowledge and institutions of education. Activists both in and out of the educational sector have built several interdisciplinary studies around countermanding the politics of exclusion from the content of education: African American studies (also called Black studies or Africana studies), ethnic studies (including but not limited to Latin American studies, Chicano studies, Native American or American Indian studies, and Judaic studies), labor studies, and, most recently, lesbian and gay studies. These growing areas all challenge the assumptions, methods, and theories of traditional disciplines by accounting for the perspectives and experiences that have, until recently, been left out of what constitutes the human experience.

Many of us trained in European-centered traditions have been persuaded to examine the content and organization of our systems of knowledge for deeply embedded cultural beliefs about inherent differences between men and women, whites and blacks, rich and poor, people who engage primarily in heterosexual activities and those who engage primarily in homosexual activities, that is, beliefs that support sexism, racism, classism, and heterosexism. Scholars across the disciplines of the humanities and social sciences have analyzed the impact of cultural beliefs about the meaning of maleness and femaleness. Termed "gender ideology"—or, more accurately, "masculinist ideology" to denote the actual power asymmetry of men and women in Western society—those beliefs predominant in the white Western world have carried implicit and explicit values about the naturalness (and correctness) of the domination of one group (male, white, propertied, colonizing) over subordinate groups (female, nonwhite, nonpropertied, colonized).

The natural sciences have received relatively less attention in feminist, African American, and similar critiques, a neglect related to the lower proportion of women and people of color in the natural sciences. Currently 16% of employed scientists and engineers are women of all backgrounds, and a total of 6% are African American, Latino, American Indian, and Pacific Islanders.[6] But even fewer scientists are actively working on gender, race, and class issues in science.

Equally important, most scientists strongly support the claim that science is relatively free of cultural and personal values, setting science apart from other systems of knowledge, and reinforcing the view that groups underrepresented in science are less interested or less talented in scientific and mathematical thinking.[7] In spite of this, feminist thinkers have identified male-centered (as well as white, heterosexist, and Eurocentric) ideologies in the organization of the scientific professions,[8] the foundations of modern science,[9] and the content of scientific knowledge, particularly in the life sciences.[10] Where the subject matter is obviously gendered, as in animal behavior studies, value-laden language and cultural beliefs about gender (associating males with activity and females with passivity, for example, or applying terms such as *harem* or *aggression* to animals other than humans) have often been superimposed onto the behaviors and social organization of animals and even plants, distorting our research programs and scientific understanding. In all areas of the sciences, we can ask: How do deeply held beliefs about gender influence the substance of science?

Significance of This Study

The significance of that question as applied to molecular biology can be appreciated on several counts. First, over the past thirty years molecular biology has moved to a dominant position in both science and society. Whether hailed as a revolution or criticized as dangerous meddling with nature, its importance was recognized early on, as the first epigraph suggests. Warren Weaver, credited with coining the term "molecular biology," was director of the Rockefeller Foundation program that determined and funded the major directions of life sciences research in the 1940s and 1950s. His view of molecular biology as "the understanding of life," a key concern of this study, is echoed by many of today's scientific leaders, as the second epigraph shows.

Indeed, molecular biology is today transforming our society's approach to disease, agriculture, pharmaceuticals, and even oil spills. Within the disciplines of the natural sciences, the dominant position of molecular biology is effecting significant changes in the boundaries of subfields in the life sciences and, thus, of systems of scientific knowledge. This is a critical moment for a feminist analysis of this field, since some leaders in molecular biology are proclaiming a reorganization of the fields of cell biology, biochemistry, and genetics, to unify them under a new molecular biology. Introduced in 1986, a leading textbook in molecular biology articulates what has been happening in biology over the past few decades:

> Biology today is scarcely recognizable as the subject that biologists knew and taught ten years ago. A decade ago, gene structure and function in the simple cells of bacteria were known in considerable detail. But now we also know that a different set of molecular rules governs gene organization and expression in all eukaryotic cells, including those of humans. . . . To comprehend fully what has been learned requires *a reformulation of a body of related information formerly classified under the separate headings of genetics, biochemistry, and cell biology. Molecular Cell Biology* aims to present the essential elements of this new biology.[11] (Emphasis added.)

At stake, then, are definitions of the central meaning of "life" and how formal science should approach the study of life at the molecular level.

Molecular biology represents a particular challenge to evaluate hidden biases about gender, sexuality, race, and class—materially significant categories of "difference" in our society—in the natural sciences because the subject matter of molecular biology (the large and small molecules of "life") is ostensibly nongendered and nonsocial. In contrast to critiques of the scientific study of organisms easily designated in our culture as male or female,[12] questions about gender beliefs in the content of supposedly nongendered topics such as biochemistry, chemistry, mathematics, geology, or physics pose an apparent problem in recognizing androcentrism, Eurocentrism, and/or racism.

My critique of an apparently gender-neutral field also carries significant implications for current debates about feminist and postmodernist transformations of epistemology.[13] Reassessments of the natural sciences exert extra weight in efforts to transform both epistemology and society, as the natural sciences function as a favored model of impartial, value-free ways of knowing. Values and epistemological issues coincide with personal issues for women's (and other subordinated groups') access to and retention in different fields in the sciences. We can ask: How do our scientific systems of knowledge frame our understanding of ourselves and our world? And who, indeed, are "we"? Efforts to make a welcome place in science for previously excluded or unwelcomed groups must identify and remove structural barriers, yet such efforts must move further to recognize and address the more subtle psychological and ethical roadblocks related to beliefs embedded, but usually denied or ignored, in the sciences.[14]

While science has the potential to improve the quality of life for many people and to contribute to liberating people from poverty, hunger, and illiteracy, recent American science has not channeled its major energies into those socially responsible projects. The issue of the uses of science affects who is attracted to do science. Thus, when we ask, for example, why there are relatively few American women in physics and engineering, we must be willing not only to address structural and psychological barriers in education and in the workplace, such as sexual harassment and employment discrimination, but also to examine the ways that those fields are directed toward military concerns, toward control over Third World resources and people, and toward consumerism. If the promises of molecular biology are primarily fulfilled in gene therapy and reproductive technologies that may benefit only a few already privileged people and harm the already disadvantaged, who will want to do that kind of science?

Critics contend that it is science, not women and other excluded groups, that must change before equitable representation can be reached.[15] And unless detrimental values and assumptions are exposed and questioned, getting more women and minorities into the natural sciences will only mean socializing and selecting for individuals from those groups who adopt the same beliefs (consciously or not) as the dominant group directing the sciences at present.

In sum, the significance of this critique of molecular biology at this time rests on the high status of science as a model of knowing, the high status of molecular biology in providing answers to major problems in society, the proposed reorganization of the science of biology around the principles and tenets of molecular biology, and the challenge to feminist analyses to deconstruct the impact of gender ideology on relatively nongendered subject matter.

We do not know enough about how ideology has affected different fields of science. This work addresses the need for case studies of the relationship of gender and science so that our theorizing is better grounded in the actual process and content of the various sciences.

Audience and Goals

This study is not intended to be comprehensive. Nor will it answer all the questions raised.[16] I hope that both its shortcomings and its strengths will serve constructively as a challenge to all those who care about molecular biology, the natural sciences, equity, women, men, and feminism to open themselves to a view of science from perspectives different from that to which they have been acculturated. As I initiated my investigations of what may appear to some to be an esoteric academic enterprise, I was encouraged by Teresa de Lauretis's "one must be willing 'to begin an argument,' and so formulate questions that will redefine the context, displace the terms of the metaphors, and make up new ones."[17]

This work aims to remedy the problem addressed by current reforms in science education. The ambitious Project 2061 of the American Association for the Advancement of Science (AAAS) is designed to generate a scientifically literate population in which scientific literacy includes "[k]nowing that science, mathematics, and technology are human enterprises and knowing what that implies about their strengths and limitations."[18] Project 2061 attempts to address much of what is left out of science education as "not science." Furthermore, the National Academy of Sciences now acknowledges that "human values cannot be eliminated from science and they can subtly influence scientific investigations," and that august body encourages scientists to address societal concerns.[19] In addition to those science educators who already incorporate societal analyses into their science teaching, I know of others who would like to incorporate such perspectives but tend to leave that task to "nonscience" courses such as women's studies, history of science, philosophy of science, ethics, and social studies of science. I hope that such educators will accept the challenges posed by this work: to raise fundamental and difficult issues of implicit and explicit values and beliefs in the processes and products of science and to create an atmosphere of questioning and, even, of discomfort in relation to science in a constructive and generative manner of inquiry to which most scientists—and feminists—aspire.

This work also attempts to broaden the democratic access to scientific knowledge in the United States, appreciating as well the significant influence of American science and technology education across the globe. As my work in Part III suggests, currently the dominant approaches and beliefs in biology and molecular biology may add subtle discriminatory pressures against certain students. Overt and covert biases and oppressive ideologies in the content of molecular biology may selectively discourage just those students from underrepresented groups (women of all racial/ethnic backgrounds and men of "minority" heritage in the U.S.), who must be tapped if we are to solve equitably the American scientific personnel shortage, including that of science teachers.

Citizens must understand science and technology to make informed choices, rather than leaving critical decisions to "a few scientists armored with a special magic."[20] Our educational system has espoused, but fallen disgracefully short of, the ideal of democratic access to scientific knowledge. For science teachers, then, our responsibility to communicate about science and to inspire *all* potential learners adds to the impetus to educate ourselves about feminist and related perspectives on the sciences. I suggest that understanding critiques of the sciences will contribute to scientific literacy in the United States. To willfully misunderstand or ignore feminist perspectives on science is to support the enslavement by sexism, racism, and classism that impoverishes the vast majority of people in the United States and, thus, society as a whole.

With the guides to writing case studies that I offer in chapter 3, I hope to stimulate people with scientific expertise and concerns (scientists and students alike) to undertake critiques of specific subfields in the natural sciences and mathematics. That includes molecular biology, since my work addresses only some of the many possible aspects of the impact of gender ideology. Such studies are needed to correct distortions in our scientific views of nature, and, thus, they should be part of the process of normal science, exposing and analyzing unwarranted assumptions. Similarly, more studies are needed that foreground race, class, sexual orientation, and other societally defined categories of "difference," such as ableness and age, that have been used to constrain human potential. Controlling for cultural biases can only promote more complete information about nature. Equally important, such studies will enrich our efforts to theorize about gender and science and thus understand how science is related to the interlocking systems of sexism, racism, heterosexism, and classism of our society. In this process we can locate the fissures, the openings to social change built into the contradictions in this complex relationship of gender and science and so chart how we can change science to play a very different role for most people.

While this work aims to effect change in the enterprise of science and science education (including that performed by journalists, teachers at all grade levels, and adults who educate through family, community, and other structures), I expect and hope that many readers will be involved in feminism, women's studies, and/or the social study of science. I particularly want to reach out to readers who want to claim or reclaim an understanding of science that may have been denied to them—and to those who want to join their knowledge and love of science with their feminism.

My Premises

Communicating across disciplinary boundaries can be difficult. The current regrettable gulf between the sciences and feminism exacerbates this problem. I draw on Francis Bacon's insight that "[e]ven to deliver and explain what I bring forward is no easy matter; for things in themselves new will yet be apprehended with reference to what is old" to caution and prepare readers for the potential confusion some may experience around terms such as *gender* and *sex, male* and *female, gene, genetics,* and *control.* Indeed, it is precisely the different, often mutually exclusive, assumptions about gender in particular that make communication between feminists and nonfeminists difficult. The meanings of those and other terms are at issue in this study of science as we address hidden values and beliefs embedded in the dominant language of sex, gender, and science. Further, I recognize that my own involvement in science, and particularly in molecular and cell biology, hinders me from breaking out of the standard language of molecular cell biology. This exercise thus becomes a step for me and for others in seeking to transcend the constraints of language laden with ideology.

Influenced in part by my experience investigating a portion of nature with scientific tools and perspectives, I start from a premise that there are things "out there" that we humans can perceive and have detected, measured, manipulated, and transformed by direct interaction or through instruments that extend human perceptions. Evidence shows, however, that our perceptions are limited, partial, and strongly shaped by our particular physicality, our conceptual frameworks, our assumptions about how things are, and who we are.[21] This understanding of science as humanly constructed knowledge is generally missing from conventional science education and practice. That is, what is termed the social or critical studies of science is not an integral part of scientists' education or self-awareness.

For me, the social construction of knowledge does not mean that society invents ideas that have no relation to physical entities and processes. I do not adhere to what is called the strong program of sociological approaches to science, in which it is charged that social factors alone determine the development of science, but I am comfortable with the concept of the social construction of science, since much evidence shows that meaning is created within a societal framework.[22] Also, I do not accept what seems to me to be an inherently classist view that language is all, or that representation supersedes materiality.

I also start from a basic and evolving premise that power and resource inequities exist in the United States and in other societies. I hold that sexism, heterosexism, racism, classism, and ethnocentrism must, for the common good, be eliminated by social change in all aspects of life. Not everyone agrees with this perspective (obviously, or society would be very different) or with what constitutes sexism and the interlinked "-isms." For example, Donna Haraway's open-ended list of the concerns of the recent feminist movement, with which I agree, includes not only reproductive rights for women, compulsory heterosexuality, and women's paid and unpaid labor, but also psychoanalysis and militarism.[23] I also start from the belief, based on accumulated evidence that the only jobs men are constitutionally incapable of performing are childbearing and wet-nursing, while the only job a woman cannot perform is sperm donation.

Foregrounding gender does not set one oppression over others. It is my view that, because systems of undeserved privilege are intermeshed and because women constitute half of every racial/ethnic group and class, an analysis foregrounding the current ideology of gender is useful to the social project of eliminating racism, hetero/sexism, and classism from this society. I embrace an analysis of society that is multifocal with regard to gender, race, ethnicity, class, and sexual orientation and that recognizes that those constraining and privileging categories may be conceptually analogous but are neither interchangeable, separable, nor additive in effect.[24] The manifestations and consequences of oppressions are experienced simultaneously by those people belonging to more than one class of oppressed categories. But *everyone* has gender, race, class, and sexual preference—not just those members of subordinated groups. White, male, upper- and middle-class, (purportedly) heterosexual—these have been the assumed universals in the last two thousand years of dominant Western thought. Analyses of privilege expose the constructed positions and values of each category, not just the subordinated ones labeled "other" relative to the supposed universal referents.

Masculine is understood here in its historically and culturally specific meaning in the tradition that has dominated the "Western world." I mean that *masculinism* embraces a belief system that is white supremacist, class-biased, Eurocentric, and heterosexist, as well as male supremacist. To my white, male readers this may sound like personal charges against entire institutions run mainly by white men, such as science. But this analysis is not intended as a blanket condemnation of all white men. Rather, feminist and related analyses offer an understanding of the *privilege* that society grants to being a white person or a male or a purported heterosexual in a society that has not eliminated racism, sexism, heterosexism, and class bias. White privilege or male privilege (or their combination) brings preferential benefits that are often invisible to those who have them, and are present even for those group members who wish to reject them. In acknowledging the kinds of privileges afforded by society to certain people and similarly acknowledging the constraints on people without those privileges, we can begin to make sense of structural, institutionalized inequities and determine how to change them to bring about equitable access to opportunities and resources. Further, we can make informed decisions about how to use the privilege we have in different situations to change society.[25]

Much of feminist thought now recognizes the race and class ideologies inherent in masculinist definitions of female and male. Groups other than white privileged people have been gendered by dominant ideology in particular ways that may contradict the ideal of "woman" or of "man."[26] Western dualistic thinking defines "the other" variously as woman, negro, Jew, homosexual, colonized, etc. And in that sense, sexism, racism, and classism have characteristics and origins in common, while the specificity of the meaning of a particular "other" at a particular moment in a particular society creates a spectrum of different experiences of oppression.

What histories of sexism and racism share, among other things, is the use of biological determinist beliefs and claims to justify the lower status of women and people of color. Nancy Leys Stepan summarizes:

> As has been well documented from the late Enlightenment on, students of human variation singled out racial differences as crucial aspects of reality, and an extensive discourse on racial inequality began to be elaborated. In the nineteenth century, as attention turned increasingly to sexual and gender differences as well, gender was found to be remarkably analogous to race, such that the scientist could use racial difference to explain gender difference, and vice versa.[27]

Likewise, heterosexism, the belief that heterosexuality is natural and universal, while homosexual behavior is a deviation from the norm, bears a particular relationship to gender ideology. The claim of natural and essential complementarity of male and female in personal, sexual, and ideological arenas renders the masculinist concept of gender heterosexist at the same time that it is sexist in defining woman only in relation to man. The compound term *hetero/sexism* is a useful reminder of the inseparable relationship of sexist and heterosexist beliefs; however, that interlocked relationship is often overlooked, and issues of sexual orientation become invisible. In addition, the relatively recent practice of condensing and reifying various sexual activities into a designation of personhood (the homosexual, the heterosexual, and, to accommodate ambiguity in a bipolar system, the bisexual) obscures the actual range of human behaviors and affectional/sexual relationships. According to a major study, at least one in five adult men in the United States has had a sexual experience with a man and about half of the men who had such experiences in the year prior to the study were or had been married to a woman.[28]

My use of gender as the focus of this analysis of molecular biology, therefore, foregrounds the sex/gender system within a matrix of systems of undeserved privilege. This foregrounding is intended as one contribution to a multifocal analysis. I have attempted also to identify and analyze where and how beliefs about race, class, and sexual orientation have impacted on this science, primarily in relation to gender ideology. The gender dualism of our society is, for me, an obvious one in the formal discourse of molecular biology, interrelated with metaphors of production and the factory. I look forward to analyses that foreground the impact of racist and similar, interlinked, and oppressive ideologies in the content of the sciences in order to eliminate them.

I offer this book particularly to those of us who are privileged to embrace both feminism and science.[29] I have learned that the privilege of access to and love of science becomes a responsibility when we also achieve a feminist consciousness.[30]

Bridging the Gulf

FEMINISM AND MOLECULAR BIOLOGY

But from the start we knew that no one had concrete
facts by which to gauge these scenarios of possible
doom [biohazards from recombinant DNA research].
So perhaps we best proceed in the fashion of the past
500 years of Western civilization, striking ahead and
only pulling back if we find the savages not of normal
size but of the King Kong variety, against which we
have no chance.
                    —James D. Watson and John Tooze, 1981[1]

[C]ertain dualisms have been persistent in Western traditions; they have all been systemic to the logics and
practices of domination of women, people of color, nature, workers, animals—in short, domination of all
constituted as others, whose task it is to mirror the self.
                    —Donna Haraway, 1985[2]

One must simply acknowledge that values do contribute to the motivations and conceptual outlook of scientists.
    —National Academy of Sciences Committee on the
                    Conduct of Science, 1989[3]

I walk a tightrope stretched across the gulf between feminism and molecular biology. On one side are scientists like James D. Watson, who speaks of conquering nature and savages for the good of society. On the other side are feminists like Donna Haraway, who deconstructs and reconstitutes the concepts—conquer, nature, society—on which Western science is based. Many more participants and positions are actually engaged, such as the National Academy of Sciences's Committee, trying to be heard by those scientists who continue to believe that science is divorced from the society that produces it. The Committee on the Conduct of Science implores us to examine the values and, with them, the biases in the construction of scientific knowledge.

The gulf between feminism and molecular biology is suggested by the relatively small number of readers who are knowledgeable about all the issues that come together in this study: what gender ideology means, what the spectrum of feminist critiques of science investigates and what attention has been given to cell and molecular biology, what molecular biology is about, and what the social studies of science and radical critiques of the use and abuse of science have addressed in relation to biology in general and molecular biology in particular. This chapter attempts to attend briefly to the problem that, for most people, the project of applying feminist perspectives to the current field of molecular biology requires introductions to systems of knowledge generally located at a distance from one another.

Scientists and others unfamiliar with gender ideology and the social construction of gender may find the first two sections of this chapter most helpful in providing background information. Readers less familiar with science will find most useful the last two sections of this chapter, on molecular biology and critiques of biological determinism. *The middle section on feminist critiques of scientific topics most relevant to my study provides information specifically related to my subsequent analyses and should be read before proceeding to Part II.*

The tightrope I (and others) walk needs many more venturesome souls, each carrying a cable across the gap, to build a bridge linking the concerns of feminist social change and those of molecular biology.

The Meaning of Gender Ideology

Western culture has long asserted that women are inferior to men and that it is in the nature of being a female to be subordinate. For example, Aristotle defined the nature of women and slaves as different from that of free men, based on the nature of the former groups to be ruled by the latter.[4] Feminists allege that such beliefs in the "naturalness" of an asymmetrical power relationship of man and woman, dominant and subordinate, constitute a key element of an ideology of gender—more precisely, a masculinist ideology—that is a foundation of sexism in our society. Such ideology, not exclusive to Western culture, justifies the power of the father, brother, husband, son, and governing fathers over women's lives.[5] In the tradition by which we are ruled in the United States, gender, like race and class, has never been the benign biological category assumed by scientists and nonscientists alike.[6]

Assertions of woman's natural inferiority and subordinate status can be traced to many of the pillars of the Western intellectual and political tradition, including Aristotle, Paul, Augustine, Thomas Aquinas, Rousseau, and even Locke.[7] It should be obvious, then, why the topic of control is of considerable interest to feminist scholars and others concerned with power relations. That Western thought is dualistic and that the predominant dualisms are arranged in a hierarchy are insights not limited to feminists, but feminists have made the connection between such binary thinking and the maintenance of unequal power relations between the sexes in society. Gender-conscious analyses have found in the major traditions of Western society the repeated theme of asymmetrical power relationships (man/woman, master/slave, man/nature) not as ancillary attitudes in society but as fundamental organizing principles for a "good society." Historian of science Donna Haraway has succinctly summarized this tenet of feminist perspectives, as quoted in the second epigraph of this chapter.[8]

With Western tradition associating reason and rule with man, and nature and subordination with woman, dichotomies central to the development of modern science are historically and culturally "genderized" (imbued with maleness and femaleness). As Sandra Harding summarizes:

> [T]he concern to define and maintain a series of rigid dichotomies in science and epistemology . . . is inextricably connected with specifically masculine—and perhaps uniquely Western and bourgeois—needs and desires. Objectivity versus subjectivity, the scientist as knowing subject versus the objects of his inquiry, reason versus the emotions, mind versus body—in each case the former has been associated with masculinity and the latter with femininity. In each case it has been claimed that human progress requires the former to achieve domination of the latter.[9]

Feminist scholars continue to investigate the validity of diverse claims of natural relationships of unequal power and the evidence and beliefs that support them. Historians and philosophers of science, L. J. Jordanova, Evelyn Fox Keller, Carolyn Merchant, and Elizabeth Potter, as well as poet Susan Griffin have explored the importance of metaphors of asymmetrical power relations, or "gender politics," in the foundations of modern science.[10] Both Keller and Jordanova highlight Francis Bacon's use of gendered political metaphors in foundational descriptions of Western science:

> My only earthly wish is . . . to stretch the deplorably narrow limits of man's dominion over the universe to their promised bounds. . . . I am come in very truth leading to you Nature with all her children to bind her to your service and make her your slave.[11]

Jordanova and Keller analyze the evolution of a male/female dualism in the prescribed relationship of the scientist to nature. At the same time, they recognize that gender-related dichotomies function in complex, historically specific ways in relation to each other.[12]

Such historical analyses of the gender-laden language of the seventeenth-century founders of modern science provide a persuasive argument for the impact of gender on the development of science. These analyses take an anti-essentialist stance that the genderization of science as a masculine endeavor and the definition of female as subordinate, emotional, nonrational, and non-scientific are historically and culturally specific creations of society.[13] Jordan-ova notes that the actual range of women's experiences, behaviors, and achievements is vastly different from the tenacious social conceptions of fixed gender attributes. This discrepancy exposes the ideological function of those dualisms in maintaining a certain social order.[14]

Motivated by explicit political concerns of social justice and equity, feminist scholarship has not only corrected distortions in many areas of knowledge but has also raised questions about the influence of cultural beliefs on episte-mology.[15] An epistemology, or system of knowledge, that validates only cer-tain ways of knowing and certain ways of transmitting knowledge functions to exclude some groups and elevate others. Rational thought, scientific method, the first knower as the most significant knower—these are desig-nated the legitimate ways of knowing. In addition, the major legitimate way of conveying knowledge is written rather than oral. And where such knowl-edge systems are coupled with systems of privilege and power, they formally and structurally limit access to training in those ways of knowing. Keller, Griffin, and historians of science Margaret Rossiter and Londa Schiebinger, although writing in very different modes, explicitly trace the masculinization of modern science to the exclusion of women from science as a profession and as an approach to knowing.[16]

The status of women has changed over time, but with a record of gains and losses that contradicts optimistic beliefs in continual progress. With organized actions from liberatory movements, opportunities and expectations for women in higher education have increased. As a result, some women with credentials and institutional positions have been able to legitimize questions about the relationship of gender and science. However, until recently, aca-demics in the history and philosophy of science, along with the predictable natural sciences themselves, have given little encouragement to their students and colleagues interested in studying women and gender in science.[17] It is clear, then, that gender-biased constraints on individuals are closely related to the production and validation of knowledge, so that feminist analyses in the natural sciences have lagged behind those in other disciplines. Further, what has been produced is often not valued as scholarship and certainly not credited as scientific scholarship, a view that only exacerbates the split be-tween the sciences and feminist critiques, as well as between science and the rest of society. Pondering epistemology from feminist and interrelated perspectives must include analyses of who has the power to create and systematize knowledge, who decides who has the power, and how gender, race, class, and other strong determinants of status in society affect the construction and choice of systems of knowledge.

Gender as Ideology: Feminist Deconstruction of Male and Female

Rejecting a belief in women as naturally inferior and subordinate to men, most feminists have theorized gender as a culturally created category, differentially contoured by race, culture, and class ("One is not born, but rather becomes, a woman").[18] While this standpoint may arise from self-interest, the evidence supporting it is vast, including the wide range of cultural meanings of male and female in different societies at different times and the evidence of differential treatment of males and females that results in different behavior. Gender—the gender dichotomy of maleness and femaleness—has been constructed as a social relationship of power (albeit derived from physical, particularly genital, differences among humans) with deeply pervasive and deeply held meanings. Following a rejection of natural difference, feminists have reexamined marriage, the family, heterosexuality, homosexuality, race relations, and occupational sex segregation as just some of the institutions that normalize and maintain a "gender ideology" of natural and necessary control of females by males.[19] Biology also comes under scrutiny.

To acknowledge the significance of ever present differences in the way humans are treated by assigned sex is to acknowledge that observed differences in behavior cannot be chalked up to "biological" differences. Adults speak to, handle, and describe babies differently, from the time of birth, based on the ascribed sex of the baby, so how can any perceived sex differences in behavior be blamed on "biology"?[20] The *perception* of biological difference and the cultural meanings attached to "male" and "female" are what condition people socially to treat boy and girl babies differently, eliciting differential responses as children grow while also socializing and conditioning the next generation. This key understanding, that biology is *inseparable* from its contextual meaning, that nature *cannot be* separated from nurture, is clearly missing from common beliefs reflected in the term "nature *versus* nurture."[21]

The concept of biology as contextual social product is equally missing from current theories of human sociobiology that assume an additive model of nature/nurture in which determinants of behavior or characteristics can be divided arithmetically. A statement, from a study discussed more fully in chapter 8, claims that seventy percent of human intelligence can be ascribed to genetic factors and thirty percent to environment.[22] That outdated and misleading model of the relationship of nature and nurture ignores the nonadditive but cumulative dynamic of biological beings interacting with their environment during both embryological development and growth and maintenance of life.

Another tenet of human sociobiology is that genetic evolution can account for behavioral and status differences between men and women; a case in point is explaining the low participation and status of women in science, business, and politics by evolutionary sex differences in "aggressivity."[23] In contrast, feminist explanations of sex differences in behavior and status draw on the lived experiences, collective and individual, of marginalized groups within the specific conditions and politics of a given moment placed in historical context.[24]

Like their biological determinist counterparts of the past, similar claims, such as that more boys than girls have "extraordinary math talent" because of inherent differences in hormones and brain structure, are deeply flawed. The concept of innate sex differences is based on highly questionable assumptions: that complex and variable behaviors can be abstracted or reified into a single identifying term such as "intelligence" or "female," and then assuming direct causal relationships to physical entities such as skull size or genes; that smaller components of matter provide more information about complex levels of organization of matter and thus can predict properties; that "causes" of behaviors can be partitioned between genetic and environmental contributing factors; and that similar behaviors are universal and, hence, biologically determined, even though exceptions can be found.[25]

It is critical to recognize that an understanding of gender as a social construct is based on a rejection of biological determinist assumptions as well as a rejection of a common form of reductionism. This critique says that male and female beings cannot be "caused" by male and female hormones or male and female genes, because universal male and female behaviors do not exist— and even if they did, that is insufficient to point to genes as causes. In addition, the fallacy of reifying complex processes into *things* is challenged when gender is understood as a cultural construct. The argument goes that "intelligence," "human nature," or "female" is each a complex of socially defined, variable behaviors, rather than a single measurable trait, and that reifying contributes to significant distortions of nature. In these ways, biological determinism, extreme reductionism, and the tendency to reify complex processes are deeply intertwined with gender beliefs.

Feminist literature on cultural biases in the sciences has addressed the propensity of scientists to impose dualistic gender and male superiority onto animals other than humans. When scientists look to nature, they often bring with them their sociopolitical beliefs about what is natural.[26] This self-reinforcing and internally consistent process creates, reflects, and reinscribes unquestioned assumptions about our world. Stereotypic attributes and behaviors, such as aggressive hunting and fighting versus passive coyness, are superimposed onto animals, often through culturally distorted language. Calling several females with a single male a "harem" conjures up quite a different power relationship from what is now called the "matriarchal" organization of elephants.

In a pertinent example, Ruth Hubbard illustrates the illogical and confusing consequences of this type of scientific sexism—the social reconstruction of maleness and femaleness—when she cites a passage on sexual behavior patterns from Wolfgang Wickler's classic book on ethology. Wickler noted the "curious" fact that among Bighorn sheep "one finds . . . no other differences whatever; the bodily form, the structure of the horns, and the color of the coat are the same for both sexes."[27] As he also observed, "the typical female behavior is absent from this pattern" and "even the males often cannot recognize a female." Furthermore, "*both* sexes play two roles, either that of the male or that of the young male. Outside the rutting season the females behave like young males, during the rutting season like aggressive older males" (Wickler's emphasis).[28]

Instead of seeing the similarities of male and female Bighorn sheep as an important lesson about species that exhibit minimal physical differences between the sexes, or sexual dimorphism[29]—as with humans—Wickler's language ("typical female behavior," "even the males") reinforces androcentrism and does so by projecting the assumptions of a heterosexist culture.

In spite of prevailing beliefs in important physical differences between the sexes, humans actually exhibit relatively little sexual dimorphism (*dimorphism* means differences between two forms; dichotomy, rather than variation, is built into the term). Most people believe that men are taller than women, but many women are as tall as or taller than many men. While it is true that the average difference in height is two to four inches, the overlap between the two populations is great, since the variation within each population (about twenty-four inches) is much greater than the average difference. It is our *customs* that create heterosexual couples in which the male, no matter how tall or short, is nearly always paired with a shorter female. The statement that "men are taller than women" is misleading and ambiguous and is based on cultural standards, not on statistical accuracy.

Another researcher's observations about the same kind of sheep also include male/female stereotyping. As the males interact, they "begin to treat each other like females and clash until one acts like a female. This is the loser in the fight." That is, the loser/female "accepts the kicks, displays, and occasional mounts of the larger without aggressive displays." Even intentionality pivots around male/female stereotyping. "The point of the fight is not to kill, maim, or even drive the rival off, but to treat him like a female."[30]

Here the sex/gender paradigm of binary maleness and femaleness is superimposed by arbitrarily designating differential behavior of males as male-like or female-like. What, indeed, do "male" and "female" mean in this view, other than winner and loser—proscribed power relations? Ruth Hubbard exposes the heterosexism embedded in this construction as well. Without the assumptions of a binary male/female sexual system or the cultural association of maleness with aggression and without the homophobia of our culture, the description above could be interpreted very differently, "say, as a homosexual encounter, a game, or a ritual dance."[31]

Another aspect of culturally loaded language of maleness and femaleness is marked by the association of (male) aggression with sexual prowess, evolutionary fitness, and achievement in general. A pointed example is taken from an issue of *Science* on sexual dimorphism in mammals, discussed in chapter 5, in which "male/female differences" are both stereotyped and justified in biological terms for the sake of "reproductive efficiency." Yet the inefficient waste of sperm produced while cyclic females are not available is explained away by linking sperm production with "the aggression necessary to fend off competitors [for] a receptive female" and by asserting that "the other aspects, such as aggression, can be useful for other life chores."[32] This cultural construction is refuted by studies of mating and social organizations in various species of animals, particularly nonhuman primates (the most aggressive male does not mate more frequently than other males), but it persists nonetheless.[33]

A strong prior commitment to asymmetrical power relations, embedded in Western thought, has distorted our knowledge of animal behavior.[34] It is logical to suspect that such associations of absence, passivity, and inferiority with femaleness may distort our understanding of the way bacteria, cells, organelles, and molecules function.

Feminist Critiques of Cell and Molecular Biology

Sandra Harding's challenging question is one guide for investigating the *current* relationship of science and gender: "How can metaphors of gender politics continue to shape the cognitive form and content of scientific theories and practices even when they are no longer overtly expressed?"[35] This question is especially relevant when considering apparently nongendered subject matter. The writings of Scott Gilbert, Emily Martin, Evelyn Fox Keller, and the Biology and Gender Study Group are particularly pertinent because they show how genderization has permeated the fields of cell and molecular biology. Recent historical studies document that scientists associated the cell's nucleus with the male and the cytoplasm with the female because a sperm cell is mainly a nucleus with a tail, while an egg cell has a nucleus and also a large volume of cytoplasm containing other cellular organelles, such as ribosomes (see figure 2-1). One consequence of this is that the term "maternal inheritance" came to be synonymous with inheritance from the cytoplasm (from material in the cytoplasm of the egg cell before fertilization that becomes active genetically after fertilization). The term is technically ambiguous since "maternal inheritance" could also refer correctly to the nuclear genetic material from the unfertilized egg (which makes up half of the nuclear genetic material after fertilization). Debates about the relationship between nu-

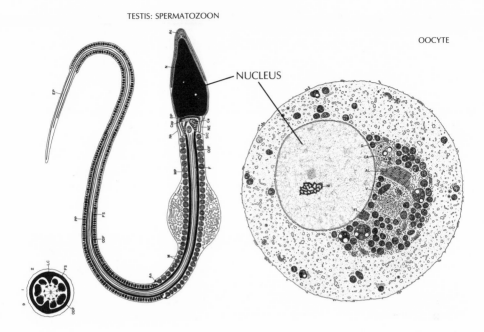

TESTIS: SPERMATOZOON

OOCYTE

NUCLEUS

Figure 2-1. Egg and sperm, illustrating the nucleus in each but differing amounts of cytoplasm. Relative sizes are not to scale. (Thomas L. Lentz, *Cell Fine Structure* [Philadelphia: W. B. Saunders, 1971], sperm, p. 247; egg, p. 269) Reproduced with permission.

cleus and cytoplasm were framed in terms of gendered power relations by using metaphors of marriage and the "appropriate" relationship between husband and wife. Several different relationships were posited, including domination of the nucleus over the cytoplasm, equal sharing of power, and domination of cytoplasm over the nucleus—each corresponding to a personal attitude related to the experience and culture of the individual (male) scientist.[36]

While it may be argued that metaphors are not the actual science, it has become increasingly obvious that metaphors and other relational aspects of language usage are the means by which data are shaped into scientific concepts (keep in mind that "data" include descriptions as well as quantitative measurements). This awareness brings into sharper focus the layers of mediation between what we call nature and what we call the science of nature: perceptions, descriptions of those perceptions, choices shaped by beliefs and efforts to make contextual meanings, interpretations, and the language used for all of these.

Both the Biology and Gender Study Group at Swarthmore and Emily Martin have traced the effects of gender ideology in descriptions of the interactions of egg and sperm cells in fertilization. The long tradition of characterizing the egg as passive and feminine and the sperm as active and masculine has changed little, with images of "the heroic sperm struggling against the hostile uterus" the typical version in biology textbooks. "The classic account . . . has emphasized the sperm's performance and relegated to the egg the supporting role of Sleeping Beauty—a dormant bride awaiting her mate's magic kiss, which instills the spirit that brings her to life." Only a few researchers "suggest the almost heretical view that sperm and egg are mutually active partners."[37]

Despite recent evidence of the egg's activity in binding and drawing in the sperm as well as blocking out extra sperm and the evidence of the very weak propulsion of sperm tails, despite evidence of the role of vaginal contractions and the sweeping action of cilia lining the fallopian tubes in moving sperm along the female reproductive tract, and even despite evidence suggesting the need for a sufficiently high sperm count to effect fertilization, the stereotype of the active individual sperm persists. Even when new evidence is discussed in terms of the active role of the egg, other culture-laden stereotypes, often negative, emerge: the nurturant female being protective, the choosy female selecting a mate, or the dangerous, spiderlike female capturing and engulfing the male.[38] Although eggs are associated by definition with the "female," we must ask if scientists are justified—and accurate—in describing them as "feminine." The consequences of cultural genderization of cells include, for science, biased hypotheses that lead to distorted research programs and, for society, inaccurate science.

Political consequences may at first seem negligible. But the power of science to define what is natural has serious repercussions when used to justify sexist views about what women can and cannot, or should and should not, do. For example, abortion rights, rights of pregnant women to their own actions, choices about fetal surgery and amniocentesis—all these are linked to the issue set up as women's rights versus fetal rights. Drawing on Rosalind Petchesky's work on representations of fetuses, Emily Martin has ferreted out the ramifications of scientific layering of intentionality, "a key aspect of personhood in our culture," onto egg and sperm, laying "the foundation for the point of viability being pushed back to the moment of fertilization." In the context of current efforts to exert further control over women's reproductive rights, Martin suggests that superimposing intentionality onto egg and sperm legitimizes the anti-choice position that fertilized eggs are little persons.[39] In this way, scientists have imbued the cell biology of fertilization with cultural beliefs about gender.

No less influential are more subtle assumptions that may act to preselect explanatory models. Evelyn Fox Keller's articulation of "prior commitments" is helpful for understanding how cultural beliefs that seem quite distant, as gender seems to be from models of biological differentiation, can and have influenced the choice of model or theory, the way a question is framed, or the interpretation given to observations.[40] Two examples from Keller's work are pertinent. The first concerns models of differentiation of slime mold cells. Drawing from her research in mathematical biology on the origin of differentiated structure, Keller analyzes competing theories about what controls cell aggregation in a slime mold. *Dictyostelium discoideum* consists of amoeba-like single cells, which, when starved, come together and form a multicellular slug that can then differentiate into a stalk and spores (see Figure 2-2). Contrary to the accepted view that signals from special cells ("pacemakers") initiate the change, the alternate model, based on Keller's and others' research, could account for the onset of aggregation in a homogeneous population of cells,

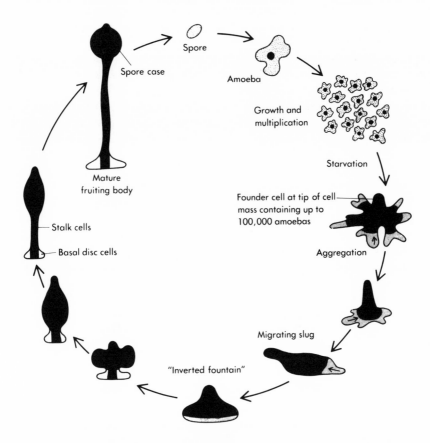

Figure 2-2. Aggregation of slime mold amoebas. "The life cycle of the cellular slime mold *Dictyostelium discoideum*. Small arrows show morphogenetic movement of cells. The first amoebas to aggregate are arbitrarily shown in the darkest color" (Watson et al., 4th ed., p. 783). Reproduced with permission.

probably by local environmental changes. Keller argues that the force of the concept of predetermined centralized control as a "natural" mode of relationship among components of living systems or populations is seen in the tenacity of one model of the origin of difference (postulating inherent superiority of special cells) over another equally plausible and supportable one (in which difference is generated by responses from all cells to environmental changes and each other). Scientists were partial to one model over the other, preferring a meaning of "difference" as intrinsically better.[41]

A second relevant example of "the truly subversive force of ideology" in science comes from Keller's biography of cytogeneticist Barbara McClintock. Addressing the multiple factors which contributed to the delay of recognition of McClintock's ultimately Nobel Prize–winning work on transposable genetic elements,[42] Keller concludes that a prior commitment to the concept of hierarchy and the domination of "master molecules," rather than interrelationships among equals, played a significant role in "the failure of the biological community to comprehend McClintock's early work on genetic transposition." Biologists, Keller charges, became "increasingly committed to the view of genes, and later of DNA, as the central actor in the cell—that which governs all other cellular processes—" while McClintock saw genes and DNA as "only one part of the cell."[43]

These two cases in cell and molecular biology, along with the ones concerning egg and sperm, demonstrate the influence of prior commitments to models of inherently centralized, hierarchical control and active/passive attributions related to gender ideology. Chapter 6 examines similar cases of unexamined assumptions. To eliminate such distortions, a new source of error—gender bias—must be acknowledged in research and thinking in biology, as Scott Gilbert and the Biology and Gender Study Group propose.[44]

To understand how gender ideology might affect molecular biology specifically, we must appreciate molecular biology as a field or territory in the enterprise of the life sciences, as specific knowledge, of course, and also as a way of thinking about life.

What Is Molecular Biology?

Simply stated, biology at the molecular level concerns the structure and function of biological macromolecules[45] and the relationship of their functioning to the structure of a cell and its internal components, including nuclei, cell membranes, and mitochondria, collectively called organelles. However, definitions of a field shift, and one aspect of this study is to highlight changes in meanings of molecular biology as a body of knowledge and as a field engaged in producing scientific knowledge.

Although molecular biology is considered a field in biology, as its history and current status demonstrate, molecular biology is now more an approach or a methodology than a distinct subdiscipline of biology. The territory of molecular biology as an academic area of science is not usually circumscribed in one department. Rather, if you look at the geography of the sciences in academe, people who teach molecular biology and do research in that inter-disciplinary area are dispersed throughout many different academic departments: biology, biochemistry, cell biology, chemistry, pathology, and physiology in medical schools, research universities, liberal arts colleges, research centers, and industrial laboratories.[46]

This is not a unique case in the biological sciences. The subfields of the study of biology (focusing here particularly at the cellular and molecular levels) overlap greatly, and territorial maps of subdisciplinary areas shift over time with changes in economic, political, and paradigmatic influences. The relationship between "fields" and "domains" in the natural sciences is a subject of interest to philosophers of science, as well as to scientists seeking financial and institutional support for their work and to students searching for an appropriate graduate department in which to train for and obtain a doctorate.

How are these distinctions presented to students? One summary of biological fields of study can be found in *Peterson's Guide to Graduate Programs in the Biological and Agricultural Sciences,* the major resource for students seeking information about graduate work in the life sciences. Appendix A[47] lists the categories in *Peterson's Guide,* showing how biology is organized into fields at this time. Biology is divided into three domains: biological sciences, agricultural sciences, and natural resource sciences. The longest list of areas is under Biological Sciences, with Biomedical Sciences tucked in there as well.

In this particular map of the biological sciences, molecular biology is linked to cell biology, while biochemistry is a separate category, and genetics is coupled with developmental biology. Those groupings are arbitrary, and the definitions given for each field are separate and distinct, even though the overlap is obvious from the field definitions (see Appendix B). Nonetheless, since *Peterson's* is a guide to graduate programs, the list is based on the institutionalization of territories of higher education within the biological and agricultural sciences and reflects a lesson to prospective students.

Appendix B includes lengthy "field definitions" that describe each area to prospective students. It is not surprising to find a significant overlap in topics in the subfields of biology. Biochemistry is described as using chemistry to understand all life processes and products. In sharp contrast, the short definitions of subfields given in the Darnell et al. textbook characterize biochemistry as focusing solely on proteins, thus presenting a much narrower view of the scope of biochemistry:

> Traditionally, the sciences of genetics, biochemistry, and cell biology—the three areas in which the greatest progress has been made in the last 25 to 30 years—used different experimental approaches and often different experimental mate-

rial. Classical geneticists sought mutations in specific genes to begin identifying the gene products and characterizing their physiological function. Biochemists tried to understand the actions of proteins, especially enzymes, from their sequences and three-dimensional structures. Cell biologists attempted to discover how specific proteins took part in the construction and operation of specialized cell structures. These subjects were taught as three courses, albeit with varying degrees of overlap.[48]

In *Peterson's Guide,* cell biology is characterized as studying how animal and plant cells "work." Molecular biology is described significantly, as starting from a focus on genetic information and then broadening to molecular organization and cell functioning. The difference in importance of these areas is suggested by the larger space given to molecular biology (twice that of either biochemistry or cell biology) and the excitement ascribed to its "offspring," genetic engineering. Most of the descriptions tend to claim breadth and interdisciplinarity as strengths. While a picture of the intertwining and practically inseparable nature of these fields emerges, several arrange themselves in a hierarchy of subdivisions; genetics is a subdivision of developmental biology, and cell biology encompasses and unites several fields, including molecular biology. Positioned within one area, not surprisingly, the scientist's view often involves elevating the role of that particular approach to biology. As chapter 7 demonstrates, leaders in molecular biology have taken that solipsistic stance to an extreme by formally proposing that all of biology be reorganized around the principles of molecular biology.

A source very different from *Peterson's Guide* provides some simple definitions of cytology (or cell biology), genetics, biochemistry, and physical chemistry (see Appendix C). These definitions, from the perspective of philosophers and historians of science, are cast around basic questions of each field and solutions to those problems. In this view: "The central problem of biochemistry is the determination of a network of interactions between the molecules of cellular species and their molecular environment . . . [while] genetics . . . has as its central problem the explanation of patterns of inheritance of characteristics."[49]

The authors are concerned with the relationship of scientific theories to institutional fields of study (and domains, which are the items of study). They also include in their delineations the *techniques* or methods that are central to each field's approach. The significance of techniques to organizing theories and subdisciplinary fields will be more fully addressed in chapter 7.

Notably, this definition of biochemistry ("the determination of a network of interactions between . . . molecules . . . and their molecular environment") is broad, like that from *Peterson's Guide* ("a study . . . to understand all life processes and the products of such processes . . . broad in its disciplinary application and . . . broad in the subjects and materials on which the scientist works"). In contrast, the description of biochemistry ("[to] understand the actions of proteins . . . from their sequences and three-dimensional structures") in the textbook *Molecular Cell Biology* is significantly more narrow. These differences reflect efforts within the "new molecular biology" to reorganize and prioritize bodies of knowledge about life at the molecular and cellular levels.

Origins of the Chemistry of Life and Issues of Reductionism and Biological Determinism

While the birth of molecular biology as a field is often dated from the development in 1953 of Watson and Crick's Nobel Prize-winning model of the structure of DNA,[50] molecular biology actually drew insights from the powerful tools of X-ray crystallography, electron microscopy, biochemical analysis of macromolecules, and molecular genetics as far back as the 1930s.[51] Molecular biology emerged as a field borrowing from, but distinct from, the existing fields of biochemistry, classical and bacterial genetics, microbiology, and physical chemistry.

Although many scientists in the 1950s and 1960s believed that the subject matter of biochemistry and molecular biology was the same,[52] the narrowing of focus and shifting of definitions in both molecular biology and biochemistry can be seen in the range of topics included by an eminent biochemist in a historical study (1800–1950) of "the interplay of chemistry and biology." Five topics were distinguished: the nature of enzymes, the chemistry of proteins, the chemical basis of heredity, the role of oxygen in biological systems, and the chemical pathways of metabolism.[53] In contrast, the focus of the field of molecular biology has been reduced to only one particular aspect of the chemistry of life: from Watson's first edition of the influential *Molecular Biology of the Gene* to the most current introductory textbooks in the field of molecular biology, "the gene" is the subject of study.[54] Not only have eminent molecular biologists singled out just *one* of the five topics in biochemistry as *the* most important, they have changed the language from the chemical *basis* of heredity to the study of how genes control the activities of life. Thus, molecular biology transformed a particular focus within biochemistry to one in which molecules other than DNA are ranked as secondary.

Tensions about the relationship between biochemistry and molecular biology, expressed by scientists such as Max Delbruck and Erwin Chargaff, were more than territorial disputes. They also reflected two different intellectual approaches to the characterization of "life" in physical terms—termed crystalline and fluid.[55] Scott Gilbert's historical study delineates the crystalline tradition as emphasizing "growth and replication as the major vital characteristics" of life. Like crystals, organisms are conceptualized in terms of their increase in size and number. In contrast, the fluid tradition focuses on "metabolism as life's prime requisite, whereby an organism retains its form and individuality despite the constant changing of its component parts," like waves or whirlpools. While both biochemistry and molecular biology claim the same goal of understanding the physicochemical basis of life, Gilbert persuasively traces biochemistry to the metabolic model of life, and molecular biology to the crystalline model via the gene.[56] By 1961, Jacob and Monod's work demonstrating a protein's modulation of gene expression brought the "rival intellectual traditions concerning the physical basis of life [together] into a scheme whereby DNA not only coded for proteins but the proteins could bind back to specific regions of DNA." Gilbert proposes that "the two traditions blended into each other, neither one predominating." My analysis of the current discourse of molecular biology suggests instead that the crystalline model has taken precedence throughout the history of the field (see chapters 6 and 7).[57]

The distinctive vision of biology created primarily by the physicists who entered and claimed the field after World War II is represented in such statements as Watson and Crick's description of their quest as a "calculated assault on the secret of life," in which "areas apparently too mysterious to be explained by physics and chemistry could in fact be so explained."[58] This "style of thought and work"[59] included a commitment to a form of reductionism most notably as it evolved in molecular genetics, the study of genetics at the molecular level in bacteria and the viruses infecting them.

Reductionism in the life sciences is inextricably linked with a mechanistic approach to nature—"the world as a clock"—built into the seventeenth-century foundations of modern science. Critiques of Cartesian reductionism recognize that assuming "the part is ontologically prior to the whole" frames problems within an ideology that predetermines the limits of our understanding of the relationship of parts and whole.[60]

As a method of investigation that involves cutting things up into smaller and smaller pieces and then reconstructing the characteristics of the system from the parts, Cartesian reductionism has been extremely successful.[61] One scientist-turned-historian noted that most biologists and chemists have not adopted an antireductionist position because scientists believe that it provides no particular advantage for the advancement of scientific knowledge, except the promise of greater accuracy.[62] In this view, reductionism wins because it works, but antireductionism might provide more accurate knowledge.

Nonetheless, critics within the sciences have objected to reductionism for a range of reasons. Most significant is the charge that Cartesian reductionism is not used only as a working philosophy but as "a commitment to how things really are," an ontological stance.[63] The problem of deriving the properties of the whole from the characteristics of the parts is illuminated by a familiar phenomenon: baking a cake. While the ingredients can be listed precisely, the cake "is not dissociable into such-or-such a percent of flour, such-or-such of butter, etc." Although each ingredient contributes to the final outcome, the cake is the result of nonadditive, complex interactions among all the components under a transformative process involving time, temperature, pan type, and altitude, not to mention the practices of the baker.[64] This analogy is particularly telling for claims that nature (the physical "ingredients") can be separated and measured apart from nurture (the complex process of baking). Ruth Hubbard notes that even the concept of interactionism between nature and nurture is inadequate to describe the nonadditive, multiplistic, nonmeasurable process; she suggests "transformationism" as a closer approximation.[65] Furthermore, as Emily Martin adds, the taste of the cake, the perception of it by each individual, is even more complex and irreducible. It is different for each occasion, for each person, for each memory it evokes.

Efforts to correlate behaviors or characteristics of living things to physical entities that determine them have serious limitations that do not arise solely from the limitations of current experimentation or instrumentation, as sociobiologists sometimes suggest, but also from the conceptual flaws of reductionism and biological determinism. Humans are not physically equipped to fly, but some societies (not all) have developed that capacity. How could that have been predicted from brain structure and function? Conversely, the finding that a medical student functioned with only ten percent of his cerebral cortex challenges assumptions about how human brains function and correlate with behavior.[66] Other factors are frequently excluded in reductionist explanations: *norms of reaction,* illustrating the unpredictable range of variation in phenotype that the same genotype can produce; *developmental noise,* biological serendipity, such as which cell ends up in one or another position during embryological development; complex *transformational dynamics,* such as stimuli and experience tangibly, but not predictably, affecting living beings at the moment of interaction and in the future as well (ingestion of milk and physical exercise during growth not only enhance bone growth but affect bone mass and decalcification in the future adult human); and *multiple nonpredictable results of such transformational dynamics* with regard to brain and nervous system in particular (stimuli alter brain structures in adult rats and mice, and human brain structures vary across different cultures).[67]

The powerful methodology of reductionism and the reductionist worldview reinforce each other, making it extremely difficult to escape their conceptual grip—or to even imagine alternative approaches that are not labeled "vitalism," a perspective ridiculed in many texts as a superstitious belief in mysterious forces of life.[68]

A reductionist perspective on biology, basing the understanding of "life" at the macromolecular level on studies of bacteria and their viruses, predominated through the 1960s, reaching an apparent peak around 1970. The conflation of reductionism as a *method* of investigation and an *explanatory stance* is a critical issue for feminist and radical science critiques.[69] The desire to take what we perceive in nature and simplify it into satisfyingly elegant formulas—such as the Central Dogma, pronounced by Crick, that the flow of information in the cell is unidirectional from the DNA to RNA to protein—transcends its purpose to instruct and becomes a "dogma" that fixes the way we think of life at the molecular level.[70] Thus, reductionism is supported and encouraged by universalizing and parsimonious tendencies in science. Single-factor models are seen as more elegant and hence more true in Western science.

Developments since 1970 appeared to turn away from some of the reductionism of the 1960s, as new techniques[71] were applied to eukaryotic organisms. New evidence showed that the genome (the entire genetic material of an organism or cell) was different in some ways in structure and function from the genome of bacteria on which molecular biology had been built. Previously ignored evidence, such as Barbara McClintock's work on transposable genetic elements, gained recognition as molecular biologists acknowledged significant diversity in the structure and function of genetic material from different types of organisms.

The identification of DNA as the genetic material and the dominant focus on heredity as the preeminent characteristic of life have added to reductionism another intertwined ideology with a long history in science: biological determinism. Whereas reductionism is not itself directly linked to gender ideology[72]—nor is it exclusive to the intellectual tradition of the crystalline model of life, as the history of the metabolic model shows—biological determinism has a long history as a popular justification for gender, race, and class inequality. Biological determinism asserts that behavior and therefore many aspects of human society are predetermined and fixed in our "biology." The biological "determinants" have differed historically with the major paradigms of scientific theories; thus, within a framework of anatomical comparisons and the importance of quantitative measurement and statistical assessment of difference, the nineteenth-century claim that brain size determined intelligence was used to "prove" that white Europeans were more intelligent than Native Americans or blacks and that men were more capable than women.

A century later, the concept of heredity became more closely fixed to genes and then to segments of DNA. The year 1969 marked a renewal of claims that biology explained societal status, with psychologist Arthur Jensen asserting that differences between whites and blacks in IQ test results were fixed in particular (albeit theoretical) genes. The history of such claims reveals a preoccupation with "explaining" only certain characteristics, such as criminality, purported sex differences, and intelligence, pointing to a clear relationship with sexist, racist, ethnocentric, and classist social projects. For example, Edward Clarke claimed that education would impair women's reproductive capacities, "evidence" used to keep women from entering Harvard and other men's colleges. The history of biological determinism is a record of gross bias in its scientific theories, methods, interpretations, and conclusions.[73]

An indirect link between reductionism and gender ideology has been proposed on the basis of the role of reductionism in biological determinist beliefs.[74] The experimentally useful reduction of genetics to genes made of DNA became the ideological basis for current biological determinism. When, in the mid-1970s, Edward O. Wilson published *Sociobiology, The New Synthesis,* he created the field of human sociobiology by extending reductionist and biological determinist approaches to primate and human behavior. Leaping from studies of insect societies to primates and thence to humans, Wilson reduced behavior to gene products and foretold a future of a "genetically accurate and hence completely fair code of ethics."[75] Wilson's claim that universal human behaviors (male dominance and aggression, female coyness, xenophobia, and altruism in kinship networks, to name a few) were genetic products of evolutionary adaptations provided a more powerful scientific argument than those previous claims by ethologists about similarities between animal and human societies.

Wilson's final chapter, "Man: From Sociobiology to Sociology," provides an example of a typical sociobiological sleight of hand: homosexual behavior (exclusively male) becomes a "role" that becomes a genetic predisposition and that is then fixed in "homosexual genes." Temporal action and social behavior become reified in Wilson's system as a fixed physical entity. Since, for example, many gay men have children and many do not assist in raising their sisters' children, Wilson's claims of "universality" of certain behaviors are ludicrous. Not surprisingly, Wilson's arguments rationalized the cultural dominance of white, heterosexual males.[76] Spurred in part by the public emergence of this new form of social Darwinism, historians of science and radical science critics have seen the relationships among reductionism, biological determinism, and hereditarianism as flawed explanatory systems and as invalid science used to underpin justifications for sexism, racism, classism, and heterosexism.[77] Such studies also documented the great harm perpetrated by biological determinist claims: denying access to higher education for white women and for women and men of color based on their purported inferiority; restricting immigration into the United States based on low scores on IQ tests given in English to people who did not speak English; sterilizing poor women, particularly African American, Native American, and Puerto Rican women; and murdering millions of Jewish people, gypsies, and homosexuals in Nazi Germany on the basis of so-called Aryan racial superiority.

In academic circles, as everywhere, values and knowledge interweave. Political stances motivate people to critique, to ignore, or to promote each new claim about biological determinism, while knowledge of the history of science in relation to racism and sexism can create justified skepticism about biological determinist claims. If that long and ongoing history of the misapplication of biological concepts does not make fair-minded people skeptical and suspicious of similar new claims, they must certainly ask themselves why.

This introductory discussion of gender ideology, reductionism, and biological determinism shows that feminist critiques, radical science critiques, and social studies of science have overlapped and are closely linked conceptually. It is important not to overlook, however, that they have for the most part been historically distinct from each other. For example, feminist perspectives are not an integral part of radical critiques of science, as this recent (1985) omission of sexism suggests: "The denial of the interpenetration of the scientist and the social is itself a political act, giving support to social structures that hide behind scientific objectivity to perpetuate dependency, exploitation, racism, elitism, colonialism."[78]

Several feminists have shown that men with leftist politics, well-meaning as many may be, have a history of ignoring women's perspectives and women's voices in discussions, meetings, conferences, and publications.[79] The record of radical science critiques and other social studies of science is improving slowly; however, even recent studies of the social, political, and economic generation of science have ignored gender ideology.[80] Therefore, without explicit reference to sexism or to gender as a key category in the matrix of social, political, and economic forces, one cannot assume that gender is taken into account.

In sum, feminist and other radical critiques have identified several common types of distortions in biology at the macroscopic level. These include, but are not limited to: creating complementary or mutually exclusive dualisms, such as nature versus nurture, homosexual versus heterosexual, and male versus female; superimposing stereotypical gender attributes onto animals and plants; constructing a natural order of hierarchies with assumptions of centralized control, casting power relationships of domination and subordination as products of evolution; and claiming that the parts determine the properties of the whole and that biology determines behavior. As we shall see, those problems are also present at the microbiological level in molecular and cell biology.

Feminists have turned their attention to the ways in which predominant cultural beliefs about maleness and femaleness have distorted knowledge about nature, particularly about women, whether in Aristotle's story of the male supplying an active principle to passive female matter, or Edward Clarke's influential claims of the negative effects of academic work on young women's capacity to reproduce (see chapter 6), or Benbow and Stanley's claims of inherent differences in extraordinary mathematical ability due theoretically to a hormone-induced feminization of women's brains (see chapter 3).[81] If the goals of scientists today include identifying and eliminating distorting biases from the content and process of science and opening science as a body of knowledge and a professional enterprise to truly represent society as a whole, then the gulf between feminism and molecular biology—and, indeed, all the natural sciences—must be bridged.

3

Methodology

Women, more than men, are bound by tradition and authority. What the father, the brother, the doctor, and the minister have said has been received undoubtingly. Until women throw off this reverence for authority, they will not develop. When they do this, when they come to truth through their investigations, when doubt leads them to discovery, the truth which they get will be theirs, and their minds will work on and on unfettered.

—Maria Mitchell, 1871[1]

The task of making science less masculine is also the task of making it more completely human.

—Londa Schiebinger, 1989[2]

Is molecular biology a gendered science? I have chosen to analyze the content of certain forms of scientific communications: textbooks and related writings by scientists for science students; the formal printed communications written by scientists for other scientists in the same field about their research; and the formal writings by scientists or science writers primarily for other scientists and science students, rather than the general public, under the auspices of a major scientific organization.

The approach I take—examining the formal language of representation of molecular biology today—is only a partial perspective, as *any* one approach would be. However, I would argue that a feminist perspective, a view from the marginalized position of women in society, is more complete and more impartial than a masculinist perspective, which must either exclude gender as a significant category of analysis or leave unquestioned the predominating cultural distortions of the meaning of gender.[3]

Scientific Journals as Sources of Representations of Molecular Biology

Scientists communicate their formal research through science journals, and two of the most important journals, *Cell* and *Science,* are the main sources for this study. Benjamin Lewin, former editor of Britain's top science journal, *Nature,* started *Cell* in 1974, and biologists recognize it as perhaps the most prestigious place to be published in molecular biology in the United States. Recently, pressure to publish quickly to get primary credit has apparently increased the dangerous tendency to sidestep peer review, one of the safeguards of objective science. As editor, Lewin exercises considerable power over determining what constitutes significant research in the field of molecular biology. Reflecting the priorities of Lewin and the other editors, *Cell*'s mini-reviews both set and promote the guiding paradigms of the field.[4] *Cell* also has the distinction of allowing, perhaps even encouraging, more dramatic imagery than is the custom in most scientific journals.[5]

Science magazine is published by the American Association for the Advancement of Science, the largest professional organization of scientists in the United States. *Science* is a major multidisciplinary science journal, unusual for including both original scientific articles and journalistic coverage of issues of policy in the scientific professions. The research reviews and reports provide rhetorical data, such as gender metaphors and prevailing and alternative paradigms, while the news coverage in each weekly issue is a rich source of information about values, assumptions, and contested areas among molecular biologists and other scientists in the United States. The scientific reports were particularly useful in determining whether and in what ways ideological traces found in textbooks are carried over in communications among scientists.

Textbooks as Primers and Critical Filters for the Next Generation

Biology textbooks construct mindsets for the next generation of scientists, physicians, and dentists. These books define the "important" questions, the framework within which these questions are addressed, and the specific knowledge required to be a functioning scientist in the field. The values and ideologies embedded in textbooks may be responsible for alienating students who, for various reasons, do not resonate to these implicit and explicit assumptions. In this way, textbooks can act as a selective filter for serious science initiates, encouraging and inspiring some while discouraging others. Disheartened students may well drop out if they do not find alternative frameworks.

I have chosen to concentrate on two widely used introductory molecular biology textbooks. James D. Watson's *Molecular Biology of the Gene,* now co-authored by other distinguished scientists, is the most recent (1987) edition of the classic text of this field.[6] And *Molecular Cell Biology,* written by James Darnell, Harvey Lodish, and David Baltimore,[7] is a comprehensive textbook that has been singled out for "setting the standard for the way the subject should be taught."[8]

One reviewer lauded this book for its chapter on the history of the development of molecular genetics, "an account of major experimental accomplishments . . . that provides a good introduction to modern molecular biology."[9] This textbook is particularly interesting because of its progressive epistemology—its attempts to place this science into a broader historical and intellectual context, to invite students to be active participants in understanding the construction of scientific knowledge, and to make molecular biology more directly relevant to people's lives.[10] This textbook can be considered, as the authors no doubt intended, as a literary embodiment of a prevailing view in the field. The excellence attributed to this textbook augments its force as a major socializing influence for the next generation of scientists and doctors.

Molecular Cell Biology was so successful that a revised edition was published just four years after the first. The two editions provide an instructive comparison: What is preserved, what is changed, and in what ways have ideological—in particular, gender—critiques affected the presentation of unifying principles in molecular biology? This last is particularly critical, since it is *Molecular Cell Biology* that proposes a "new biology" reformulated around the principles of recombinant DNA technology (see chapter 7).

The significance of my analysis rests, in part, on my claim for the authority of these textbooks. The model for this status is taken from the fields of physics and organic chemistry, where the leading college textbook authors are frequently prestigious researchers, and the textbooks organize the basic courses in that field. Indeed, the authors of *Molecular Cell Biology* explicitly compare their work to that standard.[11] Added weight comes from the stature of the textbooks' authors. James D. Watson is probably the best-known living biologist, having shared the Nobel Prize for characterizing the double-helical structure of DNA and suggesting its "self-replicating" potential from the complementarity of the two single strands that compose most DNA (see figure 3-1). Former director of Cold Spring Harbor Laboratories James D. Watson's current place in molecular biology is suggested by his recent tenure as the first head of the multibillion-dollar Human Genome Project. His coauthors are all outstanding researchers associated with major research institutions. The authors of *Molecular Cell Biology* are equally outstanding researchers in the fields of animal virology, molecular biology, and cell biology. David Baltimore is perhaps the most widely known, having shared the Nobel Prize in 1978 for detecting and characterizing the enzyme reverse transcriptase in RNA tumor viruses and demonstrating its significance in molecular mechanisms of transformation from a normal cell to a cancerous state (see figure 3-2). He was director of the Whitehead Institute for Biomedical Research at MIT and then president of Rockefeller University, perhaps one of the most prestigious academic appointments in the sciences in this country, although his term was cut short by misgivings about his involvement in an investigation of scientific fraud (see chapter 8).

Formal vs. Informal Scientific Communications

What are the limitations of depending on textbooks and scientific journals in this analysis? Latour and Woolgar, who with their anthropological approach see the scientific laboratory as "a system of literary inscription," distinguish two types of scientific literature, formal and informal. Formal communication is epitomized by the "highly structured and stylised" scientific article or paper published in a journal. Informal literature originates inside the lab and consists not only of the rough drafts, notes, and records that feed into a scientific article, but also the notations (such as radioactive count printouts, computer-generated graphs, and other "raw" data) produced with the technical apparatus of the lab.[12]

Social studies of science have raised questions about using formal scientific communications as accurate representations of how science is done and how scientific knowledge is generated, since most communication among scientists about scientific information occurs through informal channels—termed "invisible colleges"—of well-connected networks or inner circles of scientific labs.[13] For example, Susan Wright's analysis of molecular biology and safety issues around recombinant DNA technology depends on transcripts of key meetings of scientists called to address the problems of hazards of recombinant DNA research. These transcripts reveal what she characterizes as the struggle of scientists to quell public fears, what one participant called "molecular politics not molecular biology"[14] (see chapter 8).

If scientific development is more accurately reflected in those informal

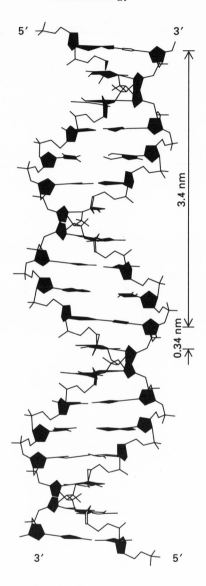

Figure 3-1. "A skeletal model of double-helical DNA," from *Molecular Cell Biology 2/E,* by Darnell, Lodish, and Baltimore, p. 69. Copyright © 1990 by Scientific American Books, Inc. Used with permission of W. H. Freeman and Company.

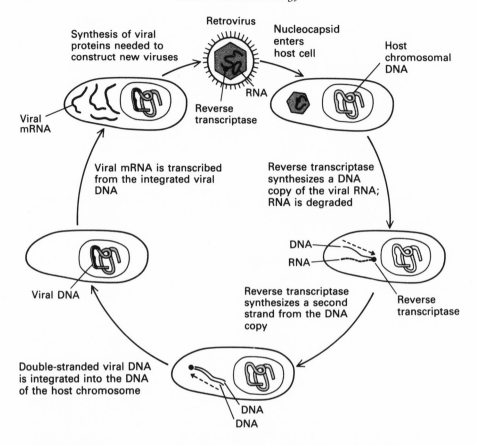

Figure 3-2. Reverse transcriptase and retrovirus transformation of cells. From
Molecular Cell Biology 2/E, by Darnell, Lodish, and Baltimore, p. 185. Copyright ©
1990 by Scientific American Books, Inc. Used with permission of W. H. Freeman
and Company.

structures and modes of communication, what, then, is the purpose of analyzing formal scientific communications and textbooks from a feminist perspective? Even with Latour and Woolgar's framework, which examines and
attempts to understand how the daily activities in the lab lead to the construction of scientific facts, published literature was found to confer legitimacy on
informal exchanges of argument and information.[15] To ignore formal scientific
communication is to miss a major point: artifacts or facets of the production
and reproduction of scientific knowledge embody particular functions and
belief systems.

The major intent of my study is to understand how the formal *representations* of molecular biology carry ideologies that may limit our understanding of nature. Because I am investigating the reproduction of knowledge in molecular biology, my work is less concerned with the intimate and candid levels of communication among scientists than with the public results codified and disseminated through scientific journal articles and texts written by respected scientists for current and future scientists. My analysis focuses on the representations of science created by scientists within the norms of formal scientific communication. The formal literature of science found in important textbooks and scientific journal articles is the site of the reproduction of information and the maintenance of norms about information content of a field, the nexus of legitimacy and communication for current and upcoming generations of scientists. While formal communications in molecular biology may not include unpublished information or the most current research or the actual processes by which scientific knowledge is produced, they are of great significance in the maintenance and reinforcement of values and ideologies among scientists and science students.

What Makes This a Feminist Inquiry?

Not every aspect of this critique of molecular biology may be easily recognized as feminist. It is appropriate, then, to ask: What distinguishes a project as feminist scholarship rather than nonfeminist or antifeminist? Is there a unique feminist method of inquiry? Sandra Harding poses these questions in her edited collection of feminist critiques of traditional social science disciplines.[16] Based on a review of some of the feminist scholarship that has challenged established beliefs in the traditional disciplines of the social sciences—rather than on abstract philosophical or other theoretical arguments—Harding concludes that there is no unique feminist method. Rather, she proposes four specific characteristics of feminist research that account for its special significance and distinguish it from conventional, nonfeminist, or antifeminist research. Feminist research originates from alternative concerns and pursues alternative purposes of inquiry, consciously chosen, as feminist awareness is chosen. In feminist research, explanatory hypotheses and evidence are not self-evident, but are recognized as politically, socially, and psychologically constructed, and a new relationship between the inquirer and her/his subject of inquiry is embraced.[17] My case study of molecular biology specifically demonstrates the four feminist characteristics discussed below.

Origins of problematics. Ideally, feminist research originates in the material and political concerns of women-centered efforts to improve the quality of life for women, children, and, hence, the planet. Women's concerns about what is wrong with society, such as violence, poverty, sexual abuse, and the misuse of power over people and resources, are placed at the center of a feminist approach to molecular biology, in contrast to conventional scientific motivations, such as the accrual of knowledge for its own sake (although the pleasure of that can be part of feminist experience), the advancement of capitalism, or personal ambition. Explicitly stated feminist values may reorder funding priorities for research, question metaphors chosen and promoted for biological life, and re-vision who should be recruited into the profession.

Purposes of inquiry. An inquiry's origin and its purpose are closely intertwined. Originating in concerns for women's quality of life, a central purpose in feminist inquiry is to eliminate constraints based on sex, race, class, sexual orientation, and other arbitrary categories, in order to change society for the better. Identifying bias in language, concepts, paradigms, applications, and personnel policies of science acknowledges the harm done to science and to society by scientific studies based on questionable assumptions about the meaning of gender, race, and sexual orientation. If one purpose of scientific inquiry is to gain more accurate understandings of the world and another is to improve the quality of life for people, then feminist inquiry, with its rigorous attention to eliminating the consequences of gender ideology, can offer corrections that expunge inaccuracies and distortions as well as encourage strict scientific standards.

Hypotheses and evidence. The discipline of biology turns to "biological" explanations first,[18] so the assumptions built into the meaning of "biology" shape what are considered plausible theories, explanations, or legitimate evidence. At this time, the culturally created meaning of the term "biological" frequently implies fixed, rigid, programmed or coded in genes—and distinguishes "biological" from "environmental factors" such as food, air, or climatic conditions. The cultural assumption that biology can be separated from environment (or "nature" from "nurture"), although it frames much of standard biological study, is illusory. What is conventionally understood as "biology" is insufficient to account for the plasticity and variety of our real biology.

Feminism turns to a broader and more interdisciplinary range of knowledges and theories for explanations and forms of evidence, taking into account cross-cultural and intracultural evidence for the plasticity of human behavior and physicality. All of this information points to a complex, nonadditive dynamic between our physical beings and the sensory and material input from our surroundings and even from our own activities.

Furthermore, the long history of erroneous and harmful theories about sex, race, and class differences creates a justifiable skepticism about hypotheses and evidence that ignore the dynamics of culture that shape both behavior and "biology." Here, politics and evidence support each other. As a conscious stance against the *status quo* that constrains certain groups of people, and in opposition to beliefs that provide little possibility for social change, feminism has analyzed the evidence of fixed determinist explanations, whether biological, psychological, or environmental.[19]

Therefore, my analysis brings a gender-conscious "control" for cultural bias in the use of gendered terms and concepts and the treatment of issues of sex and gender in molecular biology.[20] But even more profoundly, this critique exposes inaccuracies in the very definition of "biology" and requires a transformed meaning that accounts for all forms of evidence.

Relationship between the inquirer and her/his subject of inquiry. One of the tenets of feminist scholarship is that the researcher should be in the *same critical "plane"* as the subject matter. By making her conceptual framework clear, the researcher places herself on a mutual footing with the material. This helps to make visible the researcher's actual relationship to the information and interpretations of the research—and promotes a self-reflexive search for researcher bias, something that is rarely considered necessary in conventional research.[21]

An example of this self-reflexivity is found in Emily Martin's study of the ways that medical texts, current popular scientific literature, and women themselves represent the process of menstruation. Martin includes her own experience of premenstrual syndrome (PMS) and places it relative to what other women report of their own experience of menstruation. Thus, in the context of the current practice of describing PMS as a "genuine illness" and a "real physical problem," she notes that it is important for her to say what *her* experience has been with menstruation and to consider what related experiences might be relevant with regard to her own personal beliefs about PMS.[22] Imagine what would happen if all researcher/educators made visible their position relative to the subject matter. Gynecologic/obstetrics textbooks might be full of statements such as: "As a man, I have never experienced menstruation, childbirth, etc., so my reporting of what my patients have said may be affected by my lack of physical experience and empathy with these processes."

How does who I am affect my account of the subject? I have made explicit my premises and the basis for them in the first two chapters. The differences between my experience and knowledge of molecular biology and those of the scientists who have written the textbooks and articles I examine are many. But, it seems to me, it is essential that I, along with others in underrepresented groups, claim a *voice* as valid as the texts and scientific journals, in the same way that I did when I taught molecular biology and general biology.

I bring to this study a respect for the subject matter—and an expectation of reciprocity. My regard and sympathy for the field of molecular biology and its current leaders and participants come in part from my own training and participation in the field. I have come to this stance of both critic and friend by understanding my disillusionment with the idealistic and positivist ideology with which I was conditioned through the "best" education money and prestige can buy in the United States. Through my process of reevaluation, of feminist re-visioning of molecular biology, I have reclaimed my love for macromolecules, for subcellular organization, for this particular (and partial) way of understanding some of the biological aspects of nature and society and myself.[23]

It is with this mutuality, this reciprocal respect that I offer this study for science and nonscience citizens, feminist and nonfeminist alike. This critique brings a healthy scientific skepticism to the formal representations of the science of molecular biology.

Ingredients for a Feminist Critique of a Field of Science

Helen Longino and Ruth Doell have specified some "points of vulnerability in the process of science" in order to examine the "variety in the ways masculine bias can express itself in the content and processes of scientific research." Their "points of vulnerability" in the science itself include the questions addressed, the data, the hypotheses, and the distance between evidence and hypotheses. Longino and Doell apply their guidelines to the areas of human evolution and endocrinological studies of sex differences in human behavior.[24]

Drawing from these and other typologies,[25] I suggest that the following points of analysis are useful ingredients for a feminist critique of any field in the sciences and mathematics. Although not exhaustive, these facets of a full feminist critique of science must also be understood in relationship with each other, to understand the dynamics of ideology. I examine language, paradigms, unifying principles, the relationship of science to society, and the impact of gender, race, and class on who participates in shaping science.

Language is the major medium by which we communicate.[26] A feminist analysis of scientific discourse is based on the theory that language can both create and reflect—and thus perpetuate—gendered concepts that reproduce sexist, racist, and classist biases.[27] Feminist critiques of a scientific field can examine language—in textbooks, scientific literature, oral communications, popular science writing, etc.—and find evidence of the impact of gender ideology. Because manifestations of sexism are interlinked, investigation of the overt forms may reveal more subtle aspects of gender ideology and related values issues.

The use of the pseudogeneric "he" or "man" instead of "they" or "humans" (or other species of organisms), and the common cultural practice of naming an entire animal species for the male are simple, but important, examples of the role of gendered language in normalizing a cultural gender bias while obscuring the real biology of the species.[28] In spite of the notable decrease in the formal use of sexist language in many textbooks and scientific journals, overtly sexist attitudes remain part of the legacy of some science texts and some science faculty.[29] Sexual and gender metaphors are found in what could be categorized as personal narratives embedded within scientific writing, such as physicist and Nobel laureate Richard Feynman's colorful statements describing his passion for new concepts, in which he likens the new idea to a beautiful young woman, while the old theory is a sexless old woman.[30] Physics classes are notorious for faculty who use sexist slides, such as pictures of bikini-clad women to make a point about curves on a graph. Sharon Traweek has documented masculinist language and norms in the process of socializing participants in high-energy physics.[31] To diminish or ridicule women by the use of sexist language or materials is to make the classroom a chilling place for women students, as Bernice Sandler and Roberta Hall so amply document.[32]

Sexually suggestive terms, such as harem, promiscuous, coy, aggressive, and dominant, are used to describe behavior or social organization of animals, despite their obviously inappropriate anthropomorphizing. (What does "coy" or "promiscuous" mean for a mouse?) As noted in chapters 2 and 4, culturally gendered dichotomies have been superimposed not only on animals but even on plant and bacterial structures and functions. Presently, scientific credibility is undermined by projections of power inequities that model or reinforce a gender ideology of male dominance and female subordination. Thus, a key question is: *What language is used to characterize relationships among objects of study?*

Where the subject matter of the field is less obviously gendered, as in chemistry or physics or molecular biology, the language used to describe the subject matter may become less explicit in its gender symbolism.[33] Here the work of Keller, Merchant, and others[34] extends feminist critiques of science to areas of "non-sexed" subject matter. Feminist theory proposes that language founded in domination, hierarchical control, and polar dichotomies both projects and maintains masculinist ideology—a belief that white male superiority is natural and correct.

Keller also argues that the assumption that natural laws rule nature reveals "coercive, hierarchical, and centralizing" preconceptions long associated with masculine characteristics in the dominant Western society, thus making sense in the prevailing perspective.[35] In contrast, a paradigm such as that described by Barbara McClintock of a natural "order" of complex webs of interactive but nonadditive causality places beings into a relational context within their environment. In this view, hierarchical and linear cause-and-effect relationships are inadequate.[36]

Thus, attention to ideology in scientific language may reveal significant concepts that function as paradigms.

Paradigms are the frameworks or mindsets with which the scientist or scientists in a field operate.[37] Analyses of Western civilization point to the significance of paradigms that involve asymmetrical dichotomies and a deep belief in a natural order of masculine dominance and feminine subordination. We can reexamine the paradigms guiding our subfields with an awareness of the potential distortions encouraged by prior commitments to dualistic, hierarchical, and oppositional thinking.

Several paradigms may operate at the same time and be mutually reinforcing. For example, the search for biological explanations of assumed sex differences in performance on cognitive tests depends on several paradigms: behavior can be explained on the basis of biological cause-and-effect; the phenomenon of "cognition" can be reified into a single or several measurable quantities with a specific location in the body; and maleness and femaleness are fundamentally different forms of biological being. Thus, although Longino and Doell claim that "in current neuroendocrinology studies [in contrast with human evolution studies] there is no comparably explicit androcentric framework for the interpretation of data,"[38] it is the term *explicit* that is most significant. Indeed, in neuroendocrinology, the framework of maleness/femaleness is much less explicit because it functions as a buried assumption. The notion of fundamental difference between two sexes—epitomized in the naming of "male" and "female" hormones, when these are physiologically inappropriate terms—is analogous to categorizing all beings as either male or female. Thus, the use of the phrase "male and female sex hormones" both reflects and reinforces a belief in fundamental sex differences, while ultimately misrepresenting the structure and function of the hormones described. Naming certain groups of hormones "male" and "female," even though they are structurally and functionally closely related, are chemically interconverted in the human body, and are found in both male and female humans (see chapter 5), is evidence of an unsubstantiated reliance on the paradigm of bipolar sex.

When scientists raise experimental questions, they depend upon the paradigms within which they operate. Since the research hypothesis is the tentative answer to the experimental question, paradigms are intertwined inextricably with the construction and interpretation of research. For example, several tenets are embedded in the question "What are the biological explanations for sex differences in behavior?" The way the question is asked assumes that consistent and significant differences have been proven to exist, that those differences must have a biological determinant that can be detected, that biological factors in human behavior can be separated from other influences on behavior, and that studying sex differences in biological terms is fruitful for society.

The construction of experiments and research programs may be skewed as a result of unexamined paradigms. For example, using the male as the norm for sex determination has led to the neglect of developmental and evolutionary studies of the paramesonephric ducts in mammals.[39] Species chosen for extensive study tend to be those that exhibit behavior most closely fitting preconceived ideas about appropriate, "natural" behavior. Androcentric biases have also been noted in the choice of population samples, controls, re-search techniques, and data collection. Until recently, only male rats were used in learning experiments and in chemical testing. As a result, potential variations were eliminated or ignored; ironically, this tends to exaggerate some differences by minimizing subtle variations.

A unifying principle or an episteme is the overarching theory or concept that defines a field; for example, the view of the cell as the basic unit of living organisms defines cell biology. The view that social behavior is a consequence of evolution and is fixed in some way in genes operates as a unifying principle that defines the field of sociobiology. Despite sociobiologist Sarah Hrdy's cri-tique of her field for its androcentric representations of male and female pri-mates, she does not question the theory that genes control behavior and social organization.[40] In contrast, Lewontin, Hubbard and Lowe, and others have raised objections to the theory itself. My study reveals that several scientists explicitly identify a unifying principle for molecular biology and then propose using that as an overarching principle with which to reorganize all of biology. Clearly, in this case the unifying principle bears close examination.

Language, paradigms, and unifying principles represent different levels of focus for investigating cultural bias in the content of science. Shifting that focus reveals the fuzzy boundary between "good" and "bad" science.[41] While "bad" science (inaccurate or incomplete data; outright fraud) can be more clearly distinguished from "good" science at the level of construction of stud-ies (such as choice of controls and sample size), the framing of research ques-tions within a paradigm will be judged "good" or "bad" depending on factors other than those which most scientists would find scientifically sound. Thus, the field of sex differences research is "science as usual." But I would label it "bad" science (within accepted notions of scientific method) in its narrow assumptions about the causes of observed differences—more fundamentally, in its masculinist definition of gender and, most fundamentally, in its overem-phasis on difference, when nondifference is the case sometimes as much as ninety-eight percent of the time.[42] However, researchers within the field be-lieve they are adhering to scientific norms.

Interpretations and conclusions about data are shaped by paradigms and unifying principles, while the research framework influences what interpreta-tions of data are possible and desirable. A significant example involves Ca-milla Benbow and Julian Stanley's claims for the past decade that biological difference must explain at least part of the observation that as great as twelve times as many junior-high boys as girls score in the highest category on the math SAT exam. Their explanatory framework ignores at least five critical variables, including the dramatic increases in math SAT scores among girls over the past two decades and the selective pressures from parents, teachers, and peers based on cultural beliefs about math, masculinity, and femininity.

By choosing a biological determinist framework, Benbow and Stanley assume that quantitative data showing differences in performance on math tests imply inherent differences in ability, something for which there is no proof. At the same time, their conclusions are based on reducing and reifying a learned behavior into a fixed characteristic.[43] The assumptions they make about the interpretation of their data control what counts as evidence and predetermine their conclusions.

By asking how we know what we know, such investigations of cultural bias in language, paradigms, and unifying principles address *epistemological issues* as well. Epistemology has been most often the realm of philosophers of science, while scientists have not been encouraged in their training to consider questions about the assumptions within their own system of knowledge.[44]

What constitutes evidence in a particular field? What constitutes science? For example, Benbow and Stanley's work, with its tables of numbers and questionable interpretations, is reported in *Science,* while Fox and Tobin's interviews with junior-high-school boys and girls or Eccles's multifactorial studies of math performance are not.

Other issues of epistemology include how the inquirer (or "subject") is related to what is studied (the "object") in each field and what the consequences might be for different arrangements. For example, does Barbara McClintock's "feeling for the organism" have ramifications for the processes of scientific inquiry and for the uses to which she would put science that are different from those resulting from James Watson's confidence in "500 years of Western civilization, striking ahead and only pulling back if we find savages not of normal size"?[45]

How does the relationship of researcher and object of study affect the paradigms of the field, and vice versa? For example, is there a suggestive correlation between which diseases are funded for study and who the funders and scientists are? What impact do homophobia and other prejudices have on the way that the AIDS epidemic is handled? How are values embedded in the subject/object arrangement, and how do they operate?

How is objectivity related to objectification?[46] Is science distinguished from other forms of knowledge only because of the value our culture places on it? Scientists need to address these issues in a self-consciously reflexive way if science is to fulfill its potential as a system of knowledge that benefits all of humankind.

Investigating the relationship of science to society provides another analytical angle for understanding cultural bias in science. Such a critique places science in its social, political, and economic context and makes visible the applications and consequences—imagined and real—of scientific practice. In general, scientists have been socialized to demur from taking responsibility for the consequences of the work they do, pointing to politicians, corporations, the media, or the public as responsible for the distortion or misapplication of their work. Science education tends to emphasize potential benefits, without making risks and hazards visible. On the other hand, concerns about scientists' responsibilities for the uses of science are addressed directly by professional organizations such as the Federation of American Scientists, the Union of Concerned Scientists, and the Committee on Scientific Freedom and Responsibility of the AAAS.

The broad category of the social, political, and economic context of a science includes the historical legacy and origins of a field, the factors that have shaped and continue to affect a field. These dimensions are frequently the least investigated and most hidden, since science education in general does not encourage (and usually actively discourages) such views of science and since there are relatively few people—and even fewer who are feminists—doing science studies of this kind. Thus, one dimension of such an analysis is to evaluate whether and in what ways this information is hidden in the presentation of the field. To what degree and in what forums does scientific activity make this context visible? What kinds of self-awareness do scientists exhibit and communicate with regard to the social, political, and economic context of their science?[47]

The impact of gender, race, and class on who does science (and vice versa) is really a subset of the previous category, but it is of particular concern to a feminist analysis of science. A focus on gender includes the actual participation of women, their diversity with regard to race/ethnicity, class, and sexual orientation, and their location in the structure of the profession and fields as well as their visibility in textbooks and other literature.

For example, textbooks tend to credit few, if any, of the women who are major contributors to scientific fields.[48] Darnell, Lodish, and Baltimore's *Molecular Cell Biology* (second edition) is an improvement, referring to three women scientists (Martha Chase, Rosalind Franklin, and Barbara McClintock) in the introductory chapter on the history of the field. By the inclusion of her first name, Martha Chase becomes visible as a key contributor, along with Alfred Hershey, to their classic experiment, published in 1952, showing that the DNA, but not the protein, of a virus enters the bacterial cell when certain viruses initiate infection. Often referred to as the Hershey-Chase experiment, it provided significant evidence that DNA, rather than the previously suspected protein, is the genetic material.[49]

In contrast, Watson's *Molecular Biology of the Gene* is inconsistent in including scientists' first names, rather than initials. As a result, the reference to "Edmund B. Wilson and his student, N. M. Stevens" as discoverers of chromosomal determination of sex masks the identity of Nettie Maria Stevens. Graduating from Stanford University in 1899 and completing a master's degree the year after, Stevens published nine papers before completing her doctorate at Bryn Mawr College with famed geneticist Thomas Hunt Morgan. Her independent work showing that cells of males of the common mealworm have one small, unmatched chromosome (designated the Y chromosome) while cells of comparable females have paired (homologous) X chromosomes was done while she was a postdoctoral researcher at Bryn Mawr, after Morgan's departure. Watson et al.'s statement is incorrect, as Stevens was not a student of Wilson.[50]

To return to Darnell, Lodish, and Baltimore's *Molecular Cell Biology,* note that Rosalind Franklin is given equal footing with Maurice Wilkins as one of his "collaborators" in providing the X-ray diffraction patterns of DNA that Watson and Crick used to support their model of DNA as a double helix.[51] Franklin's equal credit is quite different from the inaccurate and even scornful treatment she received in James Watson's influential book, *The Double Helix.* What goes unsaid in textbooks is that Watson and Crick saw Franklin's data without her knowledge (but with the help of the institute's director, Max Perutz)—and her excellent diffraction pictures of DNA pointed to a helical structure with phosphate moieties along the outer edges, giving Watson and Crick the edge in their race for first place in determining the three-dimensional structure of DNA.[52] If biology teachers use *The Double Helix* in their classes, it is only fair to also include feminist research on Rosalind Franklin to counteract Watson's reinscription of the negative image of women scientists that may discourage women from entering science. The discovery story of the DNA double-helical structure told from the perspective of an outcast of that small scientific community can begin to convey the significance of gender, ethnicity, and race at that time and place.[53]

Barbara McClintock is the third woman mentioned in the historical section of *Molecular Cell Biology,* and here she is highlighted with a color photo and this caption: "Barbara McClintock (1902–) lecturing on her genetic work with corn, which provided the first microscopic evidence of chromosome breakage and reunion during recombination. Later McClintock showed that some chromosome segments are 'mobile' in the genome, for which she received the Nobel Prize in 1983."[54] What remains unstated is that both breakthroughs earned her the Nobel Prize and that her work on mobile genetic elements ("jumping genes") in corn was of the same significance and caliber as the work of Jacob, Monod, and Lwoff, three men rewarded with the Nobel Prize much earlier (1966) than McClintock for their elucidation of a genetic regulatory system (the operon) in bacteria. Indeed, if her genius had been recognized and rewarded earlier, McClintock's work might have been granted even greater significance because it "contradicted the prevailing dogma" in several ways: the organism (corn) that she studied was not the organism of choice (bacteria); the mode of genetic regulation (movable and nonlinear) that she found was not the assumed mechanism (nonmobile and linear); and her scientific worldview (integration and contextualism of organismic biology) was diametrically opposed to the dominant approach (reductionism and simplification of molecular genetics).[55]

The textbook chapter cited above represents women as constituting six percent of life scientists (one woman pictured among fifteen male scientists, all of whom are white; three women named out of forty-eight total). Perhaps across the history of modern science that proportion is generous, but women now constitute about twenty-five percent of life scientists, with a slightly lower percentage in molecular biology.

The 1989 edition of *Peterson's Guide to Graduate Programs in the Biological and Agricultural Sciences* makes an overtly political statement about the need for minority (but not female) participation in science, stating that existing inequities are detrimental to society in general. Concerns about gender, race, and class equity change with each edition of *Peterson's Guide;* the 1993 *Guide* made no apparent reference to women, minorities, or economic impact in science.[56] While scientific literature, such as journals with sections on news and policy issues, sometimes addresses the problem of recruiting underrepresented groups, nearly total silence reigns about the (minimally) ten percent of the scientific population who are lesbians or gay people.*

Scrutinizing scientific journals also tells us about participation in scientific research, status of participants in decision-making, and images of those participants. For example, representation of women on editorial boards of key journals in molecular biology varies widely from lows of approximately four percent to highs of about twenty-five percent, matching the rate of participation in the field.[57]

Is there a relationship between the status of women and the status of gender-related beliefs in the content and application of a field of science? While comments from socially concerned students suggest an influence, the connection is not clear, and studies in a range of fields are needed on what is often perceived as unrelated concerns. Gender bias and cultural preconceptions do not have to operate at conscious levels to create a hostile or ethically disturbing atmosphere.

This brings us full circle to the original impulse: How have the gender, race, and class of scientists and policymakers in science affected what and how science is done? Would broadening that composition change science? The record of feminist scientists and science activists points the way toward a science that would benefit more of society.

All the above ingredients are related to philosophical and epistemological concerns, including the significance of language in constructing and reflecting reality, the relationship of science to society and the problematics of the conceptual separation of science from its societal context, the interweaving of values and knowledge, and the relationship of gender, as an organizing category in society, to the organization of science.

---

*Some women identify themselves as lesbians and some as gay women. The term "lesbian" was reclaimed in affirmation in the 1970s in the women's liberation movement and continues to carry feminist political overtones.

These delineations ought to be useful, though not at all definitive, for a feminist analysis of any field of science. No one study would likely address all points. But an investigation of many of the points (and others, no doubt) should provide insights into the complex intersections of gender issues and science. For example, a mathematician using this list might find that most research articles use no apparently gendered language, while textbooks may exhibit an imbalance in gender-associated examples; male-associated examples, such as sports, might outnumber the female-associated ones, such as cooking.[58] With a consciousness about the gender associations of dichotomous, either/or thinking, hierarchies, and natural dominant/subordinate relationships or other power relations, some accepted concepts might appear arbitrary, and alternative language and concepts could be explored to create a pluralistic approach, as the new area of ethnomathematics does.[59] Further, an investigation might show that the relationship of different areas of mathematics to society is generally presented as a neutral, apolitical one or one in which progress in mathematics is linked directly to progress for all of society. In contrast, an analysis of the actual role of certain areas of mathematics might reveal important applications that affect power relations in society, for example, the development of better missile systems.

A heightened awareness of political concerns embedded in what may be frequently taught as an abstract subject might accelerate the kind of organized action taken by mathematicians in 1986–88, when President Reagan attempted to link the funding of mathematics training and research to the funding of the Star Wars project.[60] Attention to such issues would reveal the percentage of scientific research funding that is linked to potential military applications. Considerations of issues about gender, race, class, and sexuality in relation to who does science may have prompted the action taken by the American Mathematical Society and the Mathematical Association of America to cancel their 1995 joint meeting planned for Denver in response to local anti-gay-rights legislation.[61]

I do not suggest that gender and related ideologies have necessarily pervaded all aspects of the content and education of all areas of science and math, but such belief systems certainly have influenced who does science, and that relationship deserves further study. Investigations of the possible impact of gender politics are certain to promote a rethinking of the relationship of science, math, and engineering to societal values and societal priorities. I propose that as feminist perspectives on the sciences interest more and more people, opportunities and encouragement will increase for understanding scientific fields in new ways and for working on reconceptualizing the natural sciences and mathematics to eliminate gender-based and related biases.

PART II

*Gender Ideology in the Content
of Molecular Biology*

Sex and the Single Cell

DISTORTING GENETICS

Male and female *E. coli* cells are distinguished by the presence of a distinct supernumerary sex chromosome called the *F (fertility) factor.* When it is present as a discrete body, the cells are male (F+) and capable of transferring genes into female cells. In its absence, *E. coli* cells are female (F−) and act as recipients for gene transfer from male cells.

—James Watson et al., 1987[1]

Scientific theory is a particular kind of myth that answers to our practical purposes with regard to nature. It often functions as myths do, as persuasive rhetoric for moral and political purposes.

—Mary Hesse, 1989[2]

Male and Female Bacteria and the Male Signifier

"Two Distinct Sexes Are Found in *E. coli*," the authors inform readers of *Molecular Biology of the Gene;* they then explicate "sexuality" in this bacterium:

> As in higher organisms, there exist male and female cells, but these do not fuse completely, allowing their two sets of chromosomes to intermix and form two complete diploid genomes. Instead, the transfer is always unidirectional, with male chromosomal material moving into female cells; the converse movement of female genes into male cells never occurs. . . . [A]ll the cells in mixed cultures rapidly become male (F +) donor cells.[3]

This male/female designation has been customary in biology textbooks since the 1950s, when scientists found that the single-celled bacterium *E. coli*, cultivated as an experimental organism, sometimes transfers a portion of its genetic material from one cell (which in this case is a whole organism) to another. One form of genetic transfer between two cells is the movement of a tiny circle of DNA, called an F (for "fertility") plasmid, with the aid of a bridge called a pilus. The pilus is a thin protuberance that grows out from the surface of the plasmid-containing cell (called F +) and attaches to a cell without a plasmid (called F −). In this process, the plasmid replicates, and one of the resulting two copies ends up in the second cell (now F + because it contains an F plasmid), while the other copy remains in the original F + cell. (See figure 4-1.)

Scientists labeled the donor and recipient cells "male" and "female," respectively, based on the presence (male) or absence (female) of the plasmid. The language used leaves no doubt about the intended meaning of the relationship, referring to "conjugal unions between male and female cells."[4] Even though the bacteria differ only in the presence or absence of the F plasmid, scientists designate them as different "strains" of *E. coli,* exaggerating the notion of fundamental difference between males and females.

What's wrong with this designation of "male" and "female" bacteria? The scientific definition of "sex" (exchange of genetic material between organisms) is confused with two cultural meanings of "sex." The first is a sexual act between a male and a female, in which the male is the initiator who makes the sexual act happen and who donates genetic material, with the female as the passive recipient, while the second is the gender designation based on the presence of a male signifier ("his sex" has a double meaning which includes that which proves his sex/gender). The cultural meaning of "sex" as physical/genital intimacy is quite different from the scientific meaning, but scientists have not respected that distinction. Nor does this conflation coincide with the scientific definition of "male" and "female" sexes, which depends on organisms forming gametes equivalent to egg and sperm (which bacteria do not). The designation of male and female strains of *E. coli* is simply incorrect by scientific definition.

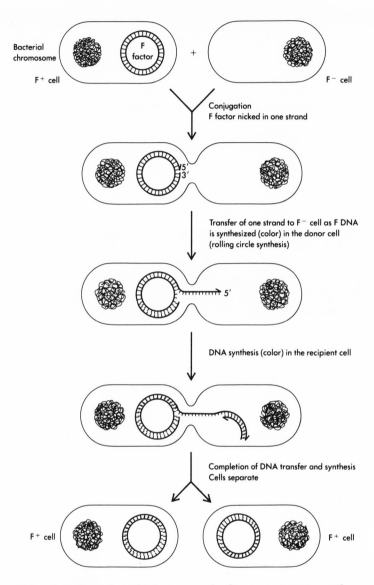

Figure 4-1. Bacteria and F factor transfer during conjugation. "The transfer of F$^+$ DNA to an F$^-$ cell. The size of the F DNA is exaggerated for clarity. One strand of the F factor is cleaved, and replication commences by a rolling circle mechanism. The newly replicated F factor passes into the F$^-$ cell at a point of direct contact between the two cells after they are linked by the F plus. Note that only the F DNA is transferred" (Watson et al., 4th ed., p. 192). Reproduced with permission.

Secondly, espousing stereotypes of the male as active and the female as passive, as well as defining female as absence, are simply sexist. Not only that, but the assumption that sexual interchange occurs only between a male and a female is a heterosexist bias. Ironically, after this "sexual" interaction involving plasmid transfer, both partners are male because they each have an F plasmid! That this "heterosexual" interaction changes the "sex" of one participant is ignored. Despite this unusual result, the authors are silent about cultural implications of this natural phenomenon for transsexuality or homosexuality.[5]

It is striking that this scientifically inaccurate and hetero/sexist sexualizing and genderizing of bacterial cells has remained largely unquestioned even in more progressive texts. The designation continues in common scientific parlance.

The issue of misleading language in biology is not new. A successful effort to eliminate an incorrect and hetero/sexist designation of male and female for a single-celled organism actually occurred in the study of eukaryotic protozoans, larger single-celled organisms such as paramecia. In the 1940s, Tracy Sonneborn convinced his colleagues that replacing the inappropriate terms *male* and *female* with nongendered terminology (*plus* and *minus, a* and *alpha*) would generate more accurate and productive research.[6]

The tenacity of genderizing nongendered beings, reflected here in the natural sciences, suggests both the power and the function of gender beliefs in our culture. Genderized attributions, even where totally inappropriate, are consistent with the worldviews of those who have the power to name and to create knowledge: in this case, scientific knowledge. It is no accident that the concept of essential male and female difference, with the male as the natural controller of action and the female defined by absence, gets embedded in the study of bacteria, so deeply held or unquestioned is our culture's belief that male/female difference is fundamental to nature. As such, it is easy to understand why such gender attributions might be unintentional and considered harmless, cute, and even useful for stirring interest in an otherwise dry subject—and why they are difficult to eliminate. Nonetheless, concerns for both accuracy and social justice should remove this ludicrous distortion.

As the Male Signifier, The Plasmid Is Dubbed an Essential Tool of Recombinant DNA Technology

While bacteria themselves are not the focus of the field of molecular biology, plasmids are central to recombinant DNA technology and molecular genetics. Some plasmids function as vectors to introduce particular pieces of DNA into bacteria or other cells for analysis. However, the F plasmid has *not* been one of the plasmids found useful in recombinant DNA technologies. Why, then, did Darnell, Lodish, and Baltimore say "these plasmids [*E. coli* F (male)] are of great use to experimental molecular biologists—they are essential tools of *recombinant DNA* technology"?[7]

With the F plasmid presented as an archetypal plasmid ("F [male] plasmid" is generalized to "these plasmids"), the male signifier is mischaracterized as an important part of recombinant DNA technology. This unintentional slip, in a textbook with few technical errors, gives the F plasmid—as the male signifier in bacteria—great importance and connects it directly to the most powerful techniques used in molecular biology, techniques proposed to unify and reorganize a "new biology."

I do not wish to make of this slip more than an unconscious expression of a deeply ingrained cultural ideology. But it is reminiscent of a similar association in a different realm of creation and creativity: literature. Sandra Gilbert and Susan Gubar asked, "Is a pen a metaphorical penis?" and found affirmatives in the words of the great white male writers "from Aristotle to Hopkins."[8] Gilbert and Gubar concluded that "[m]ale sexuality . . . is not just analogically but actually the essence of literary power. The poet's pen is in some sense (and even more than figuratively) a penis."[9]

For the scientist, it is a plasmid which signifies the phallus and is linked to tools of power. While women writers and readers of literature must contend with the problem of male control over the use and meaning of the pen(is), women scientists and others must address the problem of the male signifier associated archetypically with the foundation of DNA technologies and of the new molecular biology.[10]

The implications of a subconscious association of the bacterial plasmid with the male signifier are twofold. First, that relation suggests a powerful psychosexual factor interwoven within the high status and dominant role of recombinant DNA research (depending as it does on plasmids) in today's science and society. Secondly, guided by the analogy of male-gendering the authority of the writer, the association of the male signifier with recombinant DNA technologies reflects and sustains in yet another way the linking of scientific achievement with masculinity.

Fortunately, the second edition of this very successful textbook eliminates that statement entirely. But this example illustrates how little we may be aware of deeply held beliefs and values in this and other sciences.[11]

Distortions of Reproductive Cells and Fertilization

Active sperm and passive eggs were alive and well in the outmoded views of fertilization in the first edition of *Molecular Cell Biology*. Fertilization involves the fusion of the head (largely the nucleus) of the sperm with the entire unfertilized egg (with its nucleus and a much larger amount of cytoplasm). Just as scientists have inaccurately credited the ejaculation of sperm and the motility created by the sperm's tail with the power that propels sperm to egg, ignoring the critical role of vaginal contractions and sweeping waves of cilia lining the fallopian tubes, the textbook describes the sperm as the active agent in fertilization. The sperm "penetrates" and is "explosive"; in contrast, the egg membrane "fuses with sperm membrane," with a "depolarization of the egg plasma membrane," and a "rapid release of calcium."[12]

Contrary to the image of the sperm doing all the work by penetrating the egg surface with digestive enzymes packaged in the sperm's acrosomal cap, fusion of the egg and sperm membranes involves the activation of the sperm's enzymes by secretions from the female reproductive tract *and,* in some cases, by the protrusion from the egg's surface of microvilli that draw the sperm into the egg cell. In 1895, an eminent embryologist documented the egg's cone of microvilli, but that discovery has been virtually ignored until recently.[13]

The fusion of egg and sperm does initiate many changes, as the textbook suggests, but the egg is actively involved in ways not even hinted at. For example, microcinematography of fertilization in some species shows a startling and instantaneous change (it could be called "explosive") in the surface of the egg, involving a dramatic rearrangement of the egg cell's surface layers. Textbook language such as "the release of the calcium ions in particular is important in activating the egg for further development"[14] tends to cast the egg in a passive role, yet it is the egg that is releasing the calcium ions at its own surface!

Another kind of sexual imagery is included (in the first edition) in the description of the sperm's interaction with the egg at fertilization. The sperm cell becomes a being with a penis, in such descriptive terms as the "erection of the acrosomal process, a protrusion of the sperm" at its "penetrating" head, or "explosive polymerization."[15] The acrosomal process could just as easily be described in gender-neutral terms as a projection, like a villus. In fact, the language of the second edition is a little less phallic; the acrosome becomes a "fingerlike extension that penetrates."[16] Traces of sexualizing, such as "penetrate," remain, but the improvement is noticeable.

The influence of cultural beliefs on scientific representations of egg and sperm and related processes is not uncommon.[17] Emily Martin has documented even more subtle expressions of gender bias in physiology, in which the process of sperm development (spermatogenesis) is described in positive terms, in contrast with negative representations of egg development (ovulation). One medical textbook expresses awe: "Perhaps the most amazing characteristic of spermatogenesis is its sheer magnitude: the normal human male may manufacture several hundred million sperm per day." However, it finds no equivalent value in either menstruation or ovulation. Martin notes, "Although this text sees such massive sperm production as unabashedly positive, in fact, only about one out of every 100 billion sperm ever makes it to fertilize an egg."[18]

What difference do such distortions arising from cultural biases make for science and for society? Sexualizing sperm cells and superimposing stereo-types of the active male and the passive female distort our understanding of the process of fertilization. The harm done by inaccurately associating gender or stereotypical gender characteristics with molecules, cell organelles, or cells is not insignificant when it leads to ignoring critical data, such as the fertiliza-tion cone of certain eggs, which in turn leads to emphasizing studies of the role of sperm, but not of eggs, in fertilization.

Imposition of such ideology, unintentional though it may be, simply takes culturally pervasive but highly questionable beliefs in inherent differences be-tween men and women and embeds them at the molecular and cellular levels of our knowledge about "nature." In so doing, the power of science in general, and of this specialized and prestigious science in particular, to describe the world reinforces sexist biological determinist beliefs, despite any good inten-tions the scientists may have. Allowing tacit assumptions about the meaning of sex and gender to go unquestioned and uncorrected contributes, con-sciously or not, to the maintenance of sexism—and, by analogy, extension, and interstructuring, to the maintenance of other beliefs about inherent dif-ference.

Incorporating a feminist critique into science as an experimental control[19] could eliminate such biased language and result in more accurate descriptions of nature. Equally important, such intervention would break the destructive cycle of ideology-laden language in science reinforcing those same beliefs from which the biased language came. In this way, a feminist perspective on science can improve both scientific and equity projects and also underscore the interrelationship of science and society.

Mainstream Genetics: "Equality" Based on Overvaluing the Sperm and Undervaluing the Egg

Culturally valuing male over female at the level of cell biology results in a major distortion of the definition of genetics. One textbook's introduction to the development of "molecular cell biology" provides the usual history of ge-netics that emphasizes the *equal* contribution of egg and sperm.[20] This ac-cepted framework for genetics, based on the genetics of the nucleus, is at odds with the actual *inequality* of egg and sperm, the egg being many times larger than the sperm and contributing all the cytoplasm, including messen-ger RNA packaged in ribonucleoprotein complexes ("maternal RNA"), to the resulting fertilized egg.

The text describes the difference in size between egg and sperm but high-lights the nucleus of only the sperm: "In animals, the egg cell is comparatively large and *consists mostly of cytoplasm and stored food*. The much smaller sperm is composed *mainly of a nucleus*" (emphasis added). The logic of the inheritance of Mendelian traits—and a worldview favoring males—predominates histori-

cally: "It was reasoned that if the egg and the sperm both make equal hereditary contributions to the offspring, *the nucleus—not the cytoplasm—must hold
the key* to genetic transfer" (emphasis added).[21]

In contrast to the way it is described in the text, the egg has a nucleus of
the same size and complexity as that of the sperm. The egg also contains
nucleoprotein complexes (precursors to messenger RNA) that will provide the
fertilized egg with much of its proteins. But it is not only the maternal RNA
of the unfertilized egg that contributes significantly to the subsequent development of the zygote; all the components of the egg's cytoplasm—nutrients,
organelles such as mitochondria (which contain DNA) and ribosomes, as well
as the cell membrane and associated structures—become part of the fertilized
egg.

This great attention to genetic equality between male and female hides the
actual genetic relationship of female superiority in size of the gamete, contribution during fertilization, and involvement in embryological development.
The sex equality stressed in traditional genetics contrasts so sharply with the
cultural content of the rest of biology that it attracts a gender-conscious eye.
It is common knowledge that it is the female whose biological contribution
to progeny is greater. Many animal eggs are visible to the eye, while sperm
cells are microscopic; in fact, mature eggs are the largest cells found in animals. And, in the majority of species in which a parent physically contains the
developing progeny (exceptions include seahorses and a New Zealand frog), it
is the female that carries the developing fetus or the fertilized eggs.

Despite all this, the history of genetics unfolds as a study of the *equal* contributions of male and female to the offspring, focusing on the only component
that *is* on parity—the nucleus. The traditional emphasis on nuclear inheritance makes the great volume and important content of the egg's cytoplasm
seem irrelevant to heredity. By making nuclear heredity the organizing principle of the field of genetics, scientists have imposed onto nature the misleading
concept that egg and sperm each make "equal hereditary contributions to the
offspring." The female role is diminished by this concept of equality, reflecting
and reinforcing a belief in the relative superiority of male over female.

Sociobiologists have taken the larger size of the egg to support the notion
that females have a greater "parental investment" in their offspring (ignoring
the total volume of sperm involved in reproduction), thus determining from
"biology" that females are and ought to be the natural caretakers of progeny.
Thus, an actual *difference* in size that would, if reversed, be used to justify the
natural superiority of the larger gender, instead brings to the designated lesser
gender a "natural" responsibility for full-time care of children.

Returning to cell biology and genetics, a culturally distorted version of genetics has steered the direction of research. In the 1970s, American scientists
lowered the priority of research on the contribution of the egg's cytoplasm (in
the form of, among other things, packages of maternal RNA) and delayed
accepting the presence of DNA in nonnuclear structures in the cell, such as

mitochondria and chloroplasts.[22] Watson's textbook notes the "uneasiness" geneticists of the 1950s had with evidence that did not fit into the "neat picture of the Mendelian world with its discrete nuclear chromosomes along which genes were sited at fixed locations." The text continues, "These misbehaving genetic traits" included genetic markers that appeared to reside in the cytoplasm (termed cytoplasmic inheritance) and Barbara McClintock's transposable genetic elements, all of which were swept "under the rug . . . until the nature of more conventional chromosomal genes could be better explained."[23]

Current textbooks in molecular biology do not ignore cytoplasmic inheritance, but they accord it less importance as a topic for investigation. For example, "cytoplasmic inheritance" first appears as a subheading as far into the Darnell, Lodish, and Baltimore original edition as page 686, and is discussed in relation to mitochondrial and, later, chloroplast DNAs. In the second edition, only six and a half pages of a fifty-four-page chapter on gene control are devoted to cytoplasmic control of gene expression.[24] This imbalance comes from a self-reinforcing hierarchy of values: if inheritance is the central focus of the analysis of "life," then the nucleus will get more attention than the cytoplasm.

A report on mitochondrial DNA (mtDNA) in "research news" in *Science* supports this assessment. The article's title, "The Other Human Genome," plays with the attention given to the Human Genome Project (which focuses on nuclear DNA), a topic reported in *Science* about every third issue. In contrast, the subheading states: "Now that researchers have linked several diseases to mutations in the mitochondrial genome, mitochondrial DNA's role in other disease—even aging—is getting attention." The opening indicates the lack of attention to mtDNA until recently:

> When Doug Wallace began studying mitochondrial DNA 20 years ago, the field was something of a backwater. His was one of only a handful of groups around the world trying to figure out whether changes in these genetic oddities—packages of DNA inside a cell that are entirely separate from the chromosomes that make up the nucleus—could be responsible for human diseases.[25]

Wallace explains the difficulties of proving that mitochondrial genes are implicated in inherited diseases, since Mendelian genetics does not hold for mitochondrial DNA. Progress could be made only by "throwing away all your prejudices and starting over again."[26]

Historical studies of science tend to reproduce and amplify such imbalances within a science. Jan Sapp asserts that "the study of cytoplasmic inheritance [has] been virtually ignored by historians of modern biology."[27] One of the obvious consequences of the disproportionate attention to nuclear inheritance is the neglect of whole areas of potential importance in molecular biology (and its history, philosophy, and social studies): cytoplasmic inheritance, cytoplasmic influences on gene expression, and, more broadly, the role of the cytoplasm as a significant part of the cell. Raising this as an issue of distortions in our knowledge (and our knowledge claims) in biology highlights the insight that not all avenues or questions are or can be investigated in science, and the systematic skewing of what questions are deemed important has critical consequences for scientific knowledge and for societal uses of that knowledge.

We can only speculate about what genetics and cell biology might look like in a society of gender, race, and class equity, where our view of difference and sameness would be based on respect and appreciation, rather than on a desire to justify hierarchical rankings. In our current world, when the male can be cast in a superior light, the difference is emphasized; when the female is "better" according to the standards generally used (such as size and complexity), an effort is made to reduce the difference to some form of equality. Classical genetics in a patriarchal society has focused on the equal contributions of male and female—and thus on the nucleus and chromosomes—to obscure the greater involvement and importance of females in cellular, physiological, and hereditary phenomena. What could have been perceived as the superiority of the female has instead been exploited, for the most part, as sex difference to the advantage of the male.

Analogous claims about "equality" of men and women based on genetic parity are seen in child custody cases and the issue of parental rights to children.[28] Like the institution of law in our country, genetics embodies a liberal philosophical belief in equality that supersedes physiology and experience. The genetics of heredity has more legitimacy than the physiology of gestation and childbearing. This is reflected in a court case that granted legal custody (and legal parenthood) to the man and woman whose sperm and egg were fertilized in vitro to make an embryo, which was then implanted in the uterus of a "surrogate" mother. The woman who gestated the fetus for nine months and who delivered the baby was denied any legal status in relation to the child. The judge (Richard W. Parslow of Orange County Superior Court in Santa Ana, California) compared the woman to a foster parent who supplies shelter and food, an analogy that indeed reiterates the primacy of genetic heredity. This precedent bears out the view that our society views our genes as our "biology" and the essence of the person, establishing legal right of parenthood.[29]

This investigation started with a straightforward look at culturally gendered language and power relations in a textbook and has led to a critique of assumptions in the foundations of the field of genetics as a whole.[30] When we add up these and other examples of distorted naming, categorizing, and evaluating at the level of bacterial and animal cells, a consistent pattern emerges of a biased construction of nature that supports the dominant beliefs of white Western society. Such beliefs and the dynamic of reinforcement from science back into social consciousness and related institutions such as education, create and sustain a distorted worldview of the meaning of social relations and power relations taken from "nature."

Clearly, opportunities for change exist. Since language functions in a dynamic with societal structures and human actions, it can be a tool for change, as discussions at the end of the next chapters will illustrate. Indeed, some of the alterations made between the first and second editions of *Molecular Cell Biology* are instructive about improvements and their limitations.

This chapter highlighted the imposition of ideological male and female sex, embedded with values about male superiority, onto single-celled bacteria as well as reproductive cells in animals. The next chapter focuses on similar distortions at the level of molecules.

5

Sex and Molecules

FROM HORMONES TO BRAINS

The sex of a human or mouse embryo is normally determined by one or more genes on the Y chromosome.
                                        —Mardon et al., 1989[1]

In the presence of this gene or genes [on the Y chromosome], the bipotential gonads develop as testes, and male differentiation ensues. The absence of this gene or genes results in the development of ovaries and a female phenotype.
                                        —Mardon et al., 1989[2]

Certainly the induction of ovarian tissue is as much an active, genetically directed developmental process as is the induction of testicular tissue. . . . Almost nothing has been written about genes involved in the induction of ovarian tissue from the undifferentiated gonad.
                                        —Eicher and Washburn, 1986[3]

As we wash away the fingerprints of inaccurate notions of maleness and femaleness, what happens to cherished beliefs about our hormone molecules making us male or female? The previous chapter illustrated that language imbued with cultural gender markings distorts even molecular biology, where the subject matter is ostensibly gender-neutral. This chapter examines some of the ways that ideology-laden "sex" has affected our understanding of certain molecules and their relationship to physiology. Historical studies of biologists' construction of males and females illustrate how the power of "sex" as an unquestioned paradigm became institutionalized in the 1920s with inaccurate and misleading labeling of female and male sex hormones—and remains today in biochemistry, physiology, and molecular biology.

Sex and Molecules: The Case of Male and Female Hormones

Explicitly gendered terms are rare in molecular biology because cells, genes, and macromolecules are implicitly nongendered.[4] Nonetheless, determination of sex in humans, other animals, plants, and, as we have seen, even bacteria is a topic addressed increasingly within the framework and terminology of molecular biology, specifically, genes and gene expression. Consider the opening sentence of a research paper published in *Science* in 1989, "The sex of a human or mouse embryo is normally determined by one or more genes on the Y chromosome." First and most obviously, that statement does not make sense unless "sex" means "male sex." Yet it is commonly known that what we call the "biological sex" of a human or mouse embryo is influenced by the presence or absence of *both* X and Y chromosomes *and* autosomal genes. Conflating "sex" with "male sex" renders the female invisible. That bias then promotes a scientifically inaccurate description of development that obscures the significance of the X chromosome *and* its interactive dynamic with other chromosomes. The *Science* article reflects a predominant account—and one that is biased toward male as normative—of the development of internal reproductive organs and the prospective secondary sexual characteristics—what our society has ambiguously labeled "sex" or "gender."

Secondly, by taking as a starting point that the "sex" of humans is equivalent to the "sex" of mice and that certain genes "determine" "sex" in both species, no indication is given that "sex" in humans is actually much more than the biological assignment of male or female. This use of "sex" as "determined" by certain genes reduces the actual physico-cultural complexity of the sex/gender system to biological determinist language.

The choice of language that employs "sex" or "gender" inappropriately or without physico-cultural nuance is often the result of accepting a paradigm that has cultural flaws built into it: the unexamined paradigm of bipolar sexes. This common type of bias in the content of biology and molecular biology involves a fundamental fallacy in naming certain families of molecules "sex steroids" or "male and female" hormones. The term *sex steroids* reflects and reproduces the dominant sex/gender system, in spite of the knowledge that men and women have both "male" (androgens) and "female" (estrogens) hormones, that these male and female hormones are interconverted in our bodies, that they affect many functions besides "sex" characteristics, that men and women have differing relative amounts of these hormones, and that those relative proportions change over the life cycle (so that women after menopause have lower levels of the major estrogen and progestin than do men of the same age).[5] Diana Long Hall's historical study of endocrinology in the 1920s demonstrates that naming and characterizing certain steroid hormones as male and female sex hormones was influenced by cultural beliefs about sex differences. Further historical work by Nelly Oudshoorn documents the impact of this confusing terminology in the 1930s, when certain biological tests were adopted to standardize detection of these molecules. Not only males but also plants and fungi were found to be rich sources of "female" hormones![6]

The paradigm of biologically determined bipolar male/female "sex" has prevailed in spite of a 1939 review of a major endocrinology treatise, acknowledging that "sex" hormones are closely related and interconverted chemically and criticizing the confusion created by calling a hormone "female" when it is found in greater quantities in some males or calling a hormone "male" when it is found in ovaries. "It would appear that maleness or femaleness can not be looked upon as implying the presence of one hormone or the absence of the other, but that differences in the absolute and especially relative amount of these two kinds of substances may be expected to characterize each sex and, though much has been learned, it is only fair to state that these differences are still incompletely known."[7]

Hall's study documents as well the persistent cultural coupling of gender ideology and heterosexism when she cites one researcher's claim that men with higher levels of estrogen were "latent homosexuals."[8] Such a claim is grounded in the bipolar gender paradigm that requires the researcher to assume that, since estrogen is a "female" hormone, too much of it makes men "female." But this belief is scientifically inaccurate; estrogen is not exclusive to females, and men who engage in homosexual activities are not females. The error of assuming that male and female exist as bipolar sexes and natural sexual complements, compounded by superimposing two sexes onto certain hormones, projects ideologies of maleness and femaleness onto metabolism and other biological functions, distorting our scientific understanding of the molecular level of life.

Sexualizing Macromolecules, Like Sexualizing Bones

Scientists today continue to overlay "sex" onto molecular life. In this example from "Understanding the Bases of Sex Differences," the introductory article to an issue of *Science* devoted to sexual dimorphism (sex differences in structure), a professor of obstetrics and gynecology at Yale's School of Medicine projects "the domain of sex" onto carbohydrate metabolism by studying the actions of sex steroids: "[The authors] stress the influence of *sex steroids* on constitutive proteins (enzymes) throughout the body, *further extending the domain of sex even into processes such as carbohydrate metabolism*" (emphasis added).[9]

The errors inherent in the misapplication back in the 1920s of "male" and "female" to label closely related hormones, followed by the extension of those gender labels to chemical and physiological processes affected by the hormones, are now exacerbated in the scientific literature, including that of molecular biology. With no acknowledgment whatsoever on the part of any of the authors in that issue (or other issues) of *Science* of any problem with the term "sex steroids," readers are kept ignorant of the inaccurate and misleading labeling of "male" and "female" hormones.

Current efforts by such scientists as the one quoted above to promote the belief that "sex" extends to metabolism as well as to specific macromolecules are strikingly similar to efforts begun in Europe in the mid-eighteenth century to search for sex differences based in biology and physiology. Londa Schiebinger's historical study of human physiology and anatomy shows that drawings distinguishing between male and female skeletons first appeared between 1730 and 1790. Prior to that, in the sixteenth and seventeenth centuries, leading figures in medicine said little about sex differences in the human body. Andreas Vesalius, founder of modern anatomy, drew one human skeleton for both male and female nudes; he neither sexualized the bones, as happened later, nor believed in sex differences in the humors, as Aristotle and Galen had. (That does not mean that he or earlier scientists believed in equality. Vesalius accepted Galen's belief that women's reproductive organs were inferior because they were inverted and internal.)[10]

Schiebinger places the biology of sex differences into its historical and political context: when modern medicine rejected the principle of humors, the old explanation for difference, new explanations were needed. "Beginning in the 1750s, doctors in France and Germany called for a finer delineation of sex differences; discovering, describing, and defining sex differences in every bone, muscle, nerve, and vein of the human body became a research priority in anatomical science."[11]

She offers her study of the first representations of the female skeleton as a case study for the grander problem explored by feminists, asking, "Why does the search for sex differences become a priority of scientific research at particular times, and what political consequences have been drawn from the fact of difference? As we will see, the fact of difference was used in the eighteenth century to prescribe very different roles for men and women in the social hierarchy."[12]

Indeed, the current struggle over the meaning society makes of differences shapes the values embedded in today's studies of life at the molecular level. Sexualizing macromolecules and metabolism is as inaccurate (although in different ways)—and as political—as sexualizing bones. These examples of the impact of gender ideology on anatomy and on biochemistry and molecular biology today illustrate the way that cultural beliefs shape our understanding of biology and illuminate the process by which the meaning of maleness and femaleness are created and maintained in a society.

Determining Sex: Bipolar, Biologically Fixed, and Male-Centered

The dominant paradigm of "sex" has led to the common scientific error of equating "sex-determination" with the genetic instructions for male physiological development, specifically the development of the testes. Scientists using molecular biology remain committed to scientifically inaccurate preconceptions of sex as dualistic, biologically rigid, and favoring males. Quoting again, with emphasis added, "The *sex* of a human or mouse embryo is normally *determined* by one or more *genes on the Y chromosome*." The sentence that follows makes the assumption more evident: "In the presence of this gene or genes, the bipotential gonads develop as testes, and male differentiation ensues. The absence of this gene or genes results in the development of ovaries and a female phenotype."[13]

Describing male development as the result of the action or presence of a determinant on the Y chromosome, with female development the result of the passivity or absence of a Y-chromosome determinant, reflects a common, but male-biased, account of the development of the sexes. From her close study of androcentric bias in paradigms and descriptions of male and female development, Anne Fausto-Sterling concludes:

> . . . the supposedly general account of the development of the sexes . . . is in actuality only an account of male development. This example illustrates a case in which the meaning of *man* as a supposedly inclusive universal has slipped unnoticed into its exclusive biological category. What biologists turn out to have provided as our account of the development of gender from a mechanistic point of view is really only an account of male differentiation.[14]

The following example from a college embryology textbook suggests that this view is, indeed, the rule rather than the exception:

> The sex-determining function of the Y chromosome is intimately bound with the activity of the H-Y antigen. . . . [I]ts major function is to cause the organization of the primitive gonad into the testis. In the *absence* of the H-Y antigen the gonad later becomes transformed into the ovary. (Emphasis added by Fausto-Sterling)[15]

Another example from the same book illustrates the guiding assumption that it is the "male" hormone testosterone and other "male" secretions from the testes (the Mullerian Inhibitory Factor) that actively shape male *or* female development, while female development is mainly a result of the *absence* of "male" influence.

In a second article, "Life in the XY Corral," Fausto-Sterling offers a more extensive critique of the inaccuracies and distortions present in our current scientific view of the developmental biology of sex. (The XY Corral is the name given to the lab at MIT that was leading the search for the sex-determining gene.)[16] She cites an article published in *Cell* in which the assumption that sex-determination means male-determination guides the research program. Parenthetically, she also shows that the authors distort their data (they would say "simplify") by making it seem as if the "males" who are XX and the "females" who are XY are clearly "male" or "female," which they are not. The authors do this by omission; they refer to "sex-reversed" individuals, to "XY females" and "XX males" with no further indication or description of the individuals' characteristics. It is only in a paper authored by a different group of researchers that Fausto-Sterling finds the information that the "males" and "females" whose DNA was studied were not fertile; they lacked sperm and egg productive capacity.[17] The designation of "male" is assumed to mean the presence of the penis (even if the testes are small and nonproductive, even if the people have abnormally low testosterone and abnormally high follicle-stimulating hormone), and the designation of "female" is assumed to mean the absence of a penis—convictions common in our society. Thus, biologists' standard definition of male and female animals to mean sperm- and egg-producing types is used only in part; the presence or absence of a penis overrides other criteria.[18]

Eichler and Washburn published a critique of the scientific literature on sex determination in a respected genetics review series, summarizing a major point about the active-passive/male-female distortions in the literature:

> Some investigators have overemphasized the hypothesis that the Y chromosome is involved in testis determination by presenting the induction of testicular tissue as an active (gene-directed, dominant) event while presenting the induction of ovarian tissue as a passive (automatic) event. Certainly the induction of ovarian tissue is as much an active, genetically directed developmental process as is the induction of testicular tissue, or for that matter, the induction of any cellular differentiation process. Almost nothing has been written about genes involved in the induction of ovarian tissue from the undifferentiated gonad.[19]

In spite of the respectability of that review article (see below), published in 1986, the authors' point has been virtually ignored by researchers on sex determination at the molecular and genetic levels. It is not surprising, then, to find in 1989 and even more recently articles on the same subject that similarly ignore Eichler and Washburn's critique. My aim here is simply to note that the authors of the 1989 *Science* article, the editors of *Science,* and the peer reviewers for *Science* articles are inattentive to gender distortions in their scientific work, a stance which, if rectified, would eliminate inaccuracies in the content of science, indicate conceptual limitations in cultural constructions of science, and ultimately transform the study of biology by insisting it take into account the socially constructed nature of our view of maleness and femaleness, our perception of sex and projection of gender.

With the previous critique in mind, we can examine more closely the Mardon et al. article in *Science* on sex determination. Surprisingly, the Eicher and Washburn review *is* cited, but only with reference to a qualifying statement about the *lack* of universality of the findings reported. It seems that, in one of the two subspecies of mouse studied, "Y chromosomes appear to be ineffective in inducing testis differentiation on certain genetic backgrounds. . . . This inability may be due to allelic differences in the Y-chromosomal testis-determining genes of the two subspecies." Then, located in the penultimate paragraph of the article is the *only* reference to the involvement of the X chromosome in sex-determination (still meaning male-determination): "The availability of [certain genetic variants] would permit one to more directly examine the interactions of Y-chromosomal with autosomal or X-chromosomal sex-determining genes."[20] Within a paradigm based on cultural associations of male with activity and presence and female with passivity and absence, it is easy to understand how significant exceptions (such as a whole subspecies of mice) are cast as oddities and why research programs veer away from studies of "the interactions of Y-chromosomal with autosomal or X-chromosomal sex-determining genes" or studies of the induction of ovarian tissue.

A second article on the same topic is found adjacent to Mardon et al. in the same issue of *Science*. This article begins with more gender-neutral language: "In the fetal mouse, the paired gonadal primordia are capable of developing into either ovaries or testes." But the same biased and incorrect statement is made about the development of the female: "In the absence of a Y chromosome (for example, XX or XO karyotypes), the gonads develop into ovaries."[21] What is striking here is that the Eicher and Washburn review is the reference used for these introductory statements about sex determination! The authors do not actively take issue with Eicher and Washburn's insight about the overemphasis on female as absence; the pervasive effects of a skewed paradigm simply render it invisible.

These distortions operate significantly within molecular and genetic treatment of sex determination studies, and they permeate scientific studies about sex, differences, and sex determination.

From Molecules to Brains, Normal Science Supports Sexist Beliefs about Difference

Not only does sex determination nearly always means how *male* sex is determined, but many articles on sex determination in the past decade contain the following distortions:[22]

(1) The paradigm of sex/gender insists that subjects are either "manned" or "unmanned," polarizing male/female development into male and female beings, while categorizing physically intermediate beings as abnormal, ambiguous, or intersexed. An issue of *Science* devoted to sexual dimorphism in mammals left little room for ambiguity, though it exists both in strict biological and broader cultural terms.[23]

(2) The dominant theory guiding the relationship of sex dimorphism at the genetic and molecular level to human behavior holds that prenatal male sex hormones affect brain development and produce "sex-dimorphic" behaviors such as "energy expenditure," "social aggression," "parenting rehearsal," "peer contact," "gender role labeling," and "grooming behavior."[24] Authors make no reference to the historical and cultural overlay of bipolar sex onto hormones and minimal reference to the social construction of sex/gender.

(3) Scientists frequently make leaps between (nonhuman) animal research (for example, research on reproductive behavior such as lordosis [mounting] in rats) and implications for humans, to the advantage of the predominant theories. Or, with more subtlety, studies of rats, primates, and humans are cited *without* qualifying statements.

(4) Scientists often conflate the issue of sexual orientation with gender identity. The following illustrates the assumption that male hormones (androgens) produce a (normal male) sexual orientation to females:

> To test the validity of the prenatal hormone theory, we need to examine human subjects with endocrine disorders that involve prenatal sex-hormone abnormalities. The theory predicts that the effective *presence of androgens* in prenatal life *contributes to the development of a sexual orientation toward females,* and that a deficiency of prenatal androgens or tissue sensitivity to androgens leads to a sexual orientation toward males, *regardless of the genetic sex of the individual.*[25] (Emphasis added.)

A more recent article by those authors explicitly uses the same paradigm to explain homosexuality.[26]

The conflation, based on heterosexism, reflects and reinforces what is normal and what is abnormal in sexual relations. Obviously, in this framework, gay men are deficient in maleness and thus are more female than "normal" males; conversely, lesbians are male.

(5) Little reference is made to the social construction of sex/gender or the conflation of biological sex with gender or gender identity. Compulsory heterosexuality or the historically specific social constraints against expressing or revealing homosexual interest, behavior, or lifestyle is not mentioned. Nor are feminist and gay liberationist perspectives on the whole set of issues. For example, an article concludes:

> In the light of these generalizations, we can consider our own species. The human, like the rhesus monkey, is a species in which masculinization, rather than defeminization, appears to be the predominant mode of sexual differentiation (60). It seems reasonable that the neural substrate for gonadal steroid re-

sponsiveness is represented in the human brain in much the same way that we know it to be represented in the brains of rhesus and bonnet monkeys. . . . Other articles in this issue (87) elaborate on the extent to which we are able to recognize, *in spite of the environmental influences of learning,* the components of human behavior which are influenced by hormones during development and in adulthood.[27] (Emphasis added.)

In a very reasonable tone, a leap is made between monkeys and humans, brains and behavior. Then, too, the general scientific methodology employed can see "the environmental influences of learning" *only* as *obscuring,*[28] rather than enlightening the mechanisms of sex development, since "mechanisms" are limited by definition to "biology." And "biology" excludes the dynamic interweaving of our physical beings with our experience within our environment. Further, the articles referenced within the quote above (note 87) are the two articles in the journal that address human sexual behavior, and they are the most guilty of the inaccuracies and questionable assumptions I have delineated. In this way, mutual reinforcements at the "scientific" level render the sociopolitical construction of gender totally invisible.

While the language in the *Science* articles is carefully tempered in the tradition of scientific discourse to sound objective, the content is rife with all the problems delineated by critics—such as inadequate controls, inadequate data collection, alternative explanations, and conflicting work on prenatal hormone exposure.[29] For example, the following summary statement implies that a hormonal role in sexual orientation or cognition has been demonstrated, but not conclusively, when it has not been demonstrated *at all:* "A role of the prenatal endocrine milieu in the development of erotic partner preference, as in hetero-, homo-, or bisexual orientation, or of cognitive sex differences has not been *conclusively demonstrated"* (emphasis added).[30] Such language obscures the conclusion based on the evidence cited in the article: that there is no valid data for humans to support a role of endocrine effects, prenatal or otherwise, in sexual orientation—or, for that matter, in sex-dimorphic behavior.

Once identified, ideological distortions would seem simple to correct. Yet biological determinist views of sex differences in behavior and cognition persist. Case studies show that judgments about how much and what kind of evidence is sufficient for publication are not made solely on what scientists like to think of as scientific merit. Premature publication of questionable scientific results may not be unusual in the sciences, but *which* studies are allowed leeway and *which* studies are suppressed by not being published provide lessons in *which ideologies* are considered the norm in the biosciences.

A striking example of this has been documented by neurophysiologist Ruth Bleier, whose research on brain structure demonstrates that claims about gender-specific differences in the size of the corpus callosum in humans are inaccurate. (Note that scientists connect claims about brain differences to sex hormones with the theory that prenatal hormones masculinize or feminize the brain.) In 1982, *Science* published an article that claimed to show that the corpus callosum (a sheet of nerve fibers linking the left and right halves of the brain) was larger in human females than in males.[31] Despite their flawed methodology and statistically inadequate sample size, DeLacoste-Utamsing and Holloway applied their conclusion about morphological difference to theories of human evolution to explain purported cognitive differences and brain (cerebral) lateralization in males and females. Bleier identified several significant scientific flaws in the article—"an unstated methodology in sample selection and an unacceptable sample size, unsupportable assumptions leading to overblown interpretations, a 'finding' without a minimum standard of statistical acceptability"[32]—and launched a scientifically valid study (thirty-nine subjects as compared to fourteen; magnetic resonance images) to compare male and female brains. Bleier's results—and three more studies by other researchers—failed to find sex-related differences in the size of the corpus callosum.[33]

The Bleier group's paper was rejected by *Science*. More telling is *Science*'s rejection of a review article by Bleier, delineating the errors in methodology, conceptualization, and interpretation in several areas of sex-differences research. Although one reviewer recommended publication, the second reviewer demurred:

> While many of Bleier's points are valid, she tends to err in the opposite direction from the researchers whose results and conclusions she criticizes. While Bleier states, toward the end of her paper, that she does not "deny the possibility of biologically based structural or functional differences in the brain between women and men," she argues very strongly for the predominant role of environmental influences.[34]

As Bleier comments, the reviewer rejects the validity of such arguments for environmental influences on observed gender differences in behavior and cognition, implying that Bleier, while "erring in the opposite direction," is less legitimate and less objective than those scientists whose work *has* been published in *Science*. Clearly, the prevailing paradigm of sex differences, not balanced presentation of differing perspectives, influences publication decisions.

Feminist perspectives are often charged with being biased, because they are overtly political and come from a set of defined interests. I suggest that Bleier's case illustrates this. What is ignored is that everyone has a set of interests, but they are not usually acknowledged, particularly in the sciences, where a cult of objectivity both denies and obscures social, cultural, and economic influences.

Disagreement among scientists is not uncommon and is understood as part of an ideal of open debate of all sides of an issue. The absence from the pages of *Science* of a comprehensive critique of biological determinist theories about gender differences—and the related topic of sexuality differences—strongly suggests that one position holds sway among the decision makers. Bleier spoke about this situation at the 1987 AAAS meeting on a panel about gender issues in the sciences. A newspaper reporter quoted Holloway (one of the authors of the original study) as saying he "felt horrible" about his small sample, but "it was so intriguing we decided to publish. I didn't think it was premature at all; I felt it was damned important to get it out right away . . . [and not] wait and wait like Darwin did—and almost lose [credit for] the whole thing."[35]

Scientists Cannot Always Blame the Media

What are the consequences of having the original, scientifically question-able claim about a sex difference in the corpus callosum published in *Science,* while the scientific proof of no such difference is published in more special-ized and much less widely read journals? What are the consequences of *Science* publishing neither a visible retraction or correction to the claim nor a thor-ough critique of the field of sex differences in the brain? The January 20, 1993, issue of *Time* magazine shows how questionable scientific claims about biological determinism of sex differences in behavior and cognition reflect and reinforce prejudices about gender, with the special power of science. The magazine cover features a white boy making a muscle, while a white girl passively watches. The expression on her face conveys to me a mixture of perhaps admiration and perhaps annoyance. The public is asked "Why Are Men and Women Different?" The question itself is based on the prior assump-tion that men and women *are* different in important ways that do not have to be specified. Everyone *just knows* that the sexes are different. And the answer on the cover beams out from newsstands everywhere: "It isn't just upbring-ing. New studies show they are born that way."[36]

Under the heading, "differences that are all in the head," a picture of the brain illustrates the evidence mustered to support this assertion. Here we see the claim that the corpus callosum is "often wider in the brains of women than in those of men . . . possibly the basis for woman's intuition."[37] Six years after a spurious result is proven scientifically to be incorrect, it is presented as objective, accepted scientific fact supporting beliefs about inherent sex differ-ences. Despite the effort to cast the sex difference as a positive attribute for women, the misleading belief in inherent biologically determined differences in behavior, cognitive skills, or approaches to problem-solving gives unjusti-fiable credence to biological determinism—as well as to the view that complex processes are products of singular biological entities, whether hormones, genes, or brain cells. While the article presents a range of views on the subject of biological determinism of sex differences, it does *not* treat the inadequacies of the scientific claims themselves, leaving assumptions about scientific objec-tivity untouched.

Another questionable claim about sex differences in the brain is made about the hypothalamus, stating that a group of nerve cells was larger in heterosexual men than in heterosexual women or homosexual men. Once again, this claim is based on an article published in *Science* and has many flaws.[38] Questions about the scientific validity of this report arise from assumptions about the function of the human anterior hypothalamus that are based on nonhuman animal models; assumptions about sexual identity of sample groups; nonrandom sample; small sample size; volume versus number of cells as measure of size; effects of AIDS on brain structure; interpretation of data; and no attention given to contradictory data from cited studies.

The media are often blamed for exaggerating scientific claims or taking them out of context. Close reading of LeVay's *Science* article reveals that, while he admits that his data do not distinguish *whether the bundle of cells cause or are the consequence* of sexual orientation—and that comparisons from rats to humans in sexuality studies may not be valid—he nonetheless concludes without further justification that "it seems more likely that in humans, too, the size of INAH 3 is established early in life and later *influences sexual behavior than the reverse*" (emphasis added).[39] The abstract also states this conclusion: "This finding indicates that INAH [3] is dimorphic with sexual orientation, at least in men, and suggests that sexual orientation has a biological substrate."[40] LeVay does not (and *Science* editors did not require him to) discuss the discrepancies between his work and other research (buried in endnote 10) or the well-documented general disarray of the field of sex differences in brain studies.[41] The media cannot be blamed for taking the scientist at his or her word and trusting the peer review and editorial system to monitor scientific publication.

From Menstruation to DNA

CENTRALIZED CONTROL

[By 1952,] a small group of informed scientists knew
that DNA was the controlling molecule of life. . . . The
modern era of molecular cell biology has been mainly
concerned with how genes govern cell activity and how
proteins carry out specialized tasks.
                    —Darnell, Lodish, and Baltimore, 1990[1]

[T]hese ways of describing events are but one method
of fitting an interpretation to the facts. . . . Why not,
instead of an organization with a controller, a team
playing a game?
                    —Emily Martin, 1987[2]

What do menstruation and DNA have in common in this context? Our understanding of them is skewed by a systematic choice of a preferred model, one of centralized control within a hierarchical organization, as in the pacemaker model of slime mold differentiation discussed in chapter 2. Emily Martin's work, described in the first section of this chapter, illustrates that explanatory models of single purpose, centralized control, and natural hierarchies characterize Western physiology. Similar "prior commitments" are apparent in molecular biology in overvaluing control, heredity, and DNA, as the second section documents. Analyses of these distortions in molecular biology highlight another tendency that mimics false nature/nurture dichotomies, the misleading separation of genetic from nongenetic. A pattern of systematic skewing emerges that has consequences not only for our understanding of nature but also for our research priorities, as the final chapter of Part II will illustrate.

Menstruation and Biased Biology

Using menstruation against women has been a problem for a long time. In the 1870s, Dr. Edward Clarke published his *Sex in Education: or, A Fair Chance for Girls,* which claimed that college education would ruin women's health, particularly their reproductive health.[3] Among the responses repudiating Clarke's claims, Dr. Mary Putnam Jacobi's study of the health of college women won that foremost woman physician the 1876 Boylston Essay Prize from (ironically) Harvard College, the alma mater Clarke was trying to protect from the presence of women. One thread of the second wave of the women's movement critiqued cultural stereotypes in research and in medical texts that characterized menstruation as an illness and menopause as a pathological condition of hormone deficiency. With a pro-woman view, previously unexamined biases were exposed, revealing poor research designs and overtly negative attitudes toward women's reproductive biology and experience.[4]

Emily Martin's analysis of contemporary medical discourse has extended such investigations to identify guiding paradigms embedded in the discourse of medical texts. In *The Woman in the Body: A Cultural Analysis of Reproduction,* Martin delineates several assumptions which undergird Western medicine's "understanding" of women's physiology. One of these is that the single purpose of the process of menstruation is reproduction; within that single-purpose framework, then, the process of menstruation is the sign of failure to get pregnant. Another paradigm she identifies is a relationship of hierarchical, centralized control among the parts of the body (the hypothalamus, pituitary, ovaries, etc.) in which the brain is perceived as issuing commands to passively responsive organs.

Scientists draw paradigms or implicit rules from a culture that is negative toward women. As a consequence, textbooks represent the physiological changes in the uterus during menstruation both as negative processes of deterioration and repair that involve increased weakness, and as responses by a subservient organ to an executive brain. Martin proposes that the language in those texts both distorts and restricts our ways of conceptualizing these processes.

To illustrate her analysis, Martin reproduces the following selection from a human physiology textbook portraying a woman's failure to become pregnant as producing physiological decline:

> If fertilization and pregnancy do not occur, the corpus luteum degenerates and the levels of estrogens and progesterone decline. As the levels of these hormones decrease and their stimulatory effects are withdrawn, blood vessels of the endometrium undergo prolonged spasms (contractions) that reduce the bloodflow to the area of the endometrium supplied by the vessels. The resulting lack of blood causes the tissues of the affected region to degenerate. After some time, the vessels relax, which allows blood to flow through them again. However, capillaries in the area have become so weakened that blood leaks through them. This blood and the deteriorating endometrial tissues are discharged from the uterus as the menstrual flow. As a new ovarian cycle begins and the level of estrogens rises, the functional layer of the endometrium undergoes repair and once again begins to proliferate.[5]

Martin highlights the succession of terms confronting the reader: " 'degenerate', 'decline', 'withdrawn', 'spasms', 'lack', 'degenerate', 'weakened', 'leak', 'deterioration', 'discharge', and after all that, 'repair'."[6]

Quoting another textbook and a diagram of changes in the endometrium during the menstrual cycle, she points to the "imagery of catastrophic disintegration: 'ceasing', 'dying', 'losing', 'denuding', and 'expelling' " frequently found in many texts and conveying "failure and dissolution." She summarizes: "[U]nacknowledged cultural attitudes can seep into scientific writing through evaluative words."[7]

Comparisons with similar processes, such as spermatogenesis and stomach lining functioning, highlight the uniquely negative treatment of menstruation and the assumption that menstruation should be defined by its goal of reproduction. While metaphors of degeneration, weakening, deterioration, and repair abound in descriptions of menstruation, language describing the functioning of the lining of the stomach reflects more positive concepts of secretion, protection, and periodic renewal. Martin summarizes:

> The lining of the stomach must protect itself against being digested by the hydrochloric acid produced in digestion. In the several texts quoted above, emphasis is on the *secretion* of mucus, the *barrier* that mucous cells present to stomach acid, and—in a phrase that gives the story away—the periodic *renewal* of the lining of the stomach. There is no reference to degenerating, weakening, and deteriorating, or repair, or even the more neutral shedding, sloughing, or replacement.[8]

A major principle Martin identifies is the control of the body, by the brain, monitored through an information system:

> Although there is increasing attention to describing physiological processes as positive and negative feedback loops so that like a thermostat system no single element has preeminent control over any other, most descriptions of specific processes give preeminent control to the brain. . . . [T]he female brain-hormone-ovary system is usually described . . . as a hierarchy, in which the "directions" or "orders" of one element dominate . . .[9]

Martin's work provides a model case study for the subtle influence of cultural beliefs on the representation of biological processes. Martin proposes that conventional views constrain our thinking about women's physiology, as she suggests alternative conceptualizations:

> I have presented the underlying metaphors contained in medical descriptions of menopause and menstruation to show that these ways of describing events are but one method of fitting an interpretation to the facts. . . . Would it be . . . possible to change the nature of the relationships assumed . . . ? Why not, instead of an organization with a controller, a team playing a game? . . . Eliminating the hierarchical organization and the idea of a single purpose to the menstrual cycle also greatly enlarges the ways we could think of menopause.[10]

While human physiology is, of course, associated with gendered humans, the *actual* subject matter under scrutiny includes physiological processes and body parts such as glands that themselves are no more physically gendered as tissues and organs than a woman's stomach is. Cultural biases nonetheless operate in these descriptions, as we find negative values embedded in the language used to describe menstruation as compared to the functioning of the stomach lining or other tissues, suggesting subtle ways in which those culturally constructed attitudes function and are maintained. These distortions function to reinforce beliefs in a hierarchy of value, masculine superiority, and "difference" as "better or worse than." Martin's suggestions of alternative paradigms to hierarchy, centralized control, and single causes helps us to understand the conceptual limitations about physiology created by such distortions.

In sum, Emily Martin's discourse analysis of medical texts identifies several themes commonly found in the sciences (as in other aspects of society): negative associations with female activities, characteristics, or bodily functions; assumptions of a natural order of centralized control and hierarchical relationships, with values placed at the top of the hierarchy; and a model from modern industrialism of production as a fundamental goal of all systems—all of which create a filter in the lens through which we perceive nature. Such distortions are inappropriate, inaccurate, and also dangerous to humans in supporting oppressive beliefs.

Similar tacit assumptions (single purpose, centralized control, and hierarchies) about the natural organization and function of biological components emerge in an analysis of the discourse in molecular biology. Feminist science critics have questioned singling out DNA as the "master molecule" in the foundations of molecular biology through the 1950s and 1960s (see chapter 2). Now we can ask if the language describing the fundamental concepts of the field remains one of control by DNA and genes over everything else in the cell. Or have the insights from the more recent study of organisms other than bacteria modified the central paradigm of hierarchical control in the cell?

The Central Paradigm of Molecular Biology Today

What scientists sometimes define as the study of interactions of many different kinds of molecules and organelles is currently understood as being primarily about genes controlling what goes on (the chemical and physicochemical activities) in cells. The beginning of the preface of the newest edition of James Watson's *Molecular Biology of the Gene,* until recently the leading textbook in the field, assumes, rather than proposes, one particular meaning, focus, and significance of molecular biology: "Today no molecular biologist knows all the important *facts about the gene.* . . . It is only in this fourth edition that we see the extraordinary fruits of the recombinant DNA revolution. Hardly any contemporary experiment on *gene structure or function* is done today without recourse to ever more powerful methods for cloning and sequencing genes." The authors baldly assert that the molecular biology in the book "is by any measure an extraordinary example of human achievement." Furthermore, the amount of critical information about DNA is huge: "DNA can no longer be portrayed with the grandeur it deserves in a handy volume that would be pleasant to carry across campus." To novices, "this book [is] their first real introduction to *gene structure and function*" (emphasis added).[11]

The other major textbook in molecular biology provides the same definition: "Successful studies of the structure of genes and the control of protein synthesis by genes form the recent history of molecular cell biology."[12]

According to these texts, molecular biology today is about "gene structure and function," "facts about the gene," and "the control of protein synthesis by genes." This view, obvious and simple as it may sound to people in that field, nonetheless obscures corollary assumptions that define molecular biology as something quite different from biochemistry.

The definition of molecular biology as synonymous with or dominated by gene structure and function is corroborated by the division of topics in journals, as well as by the ordinary language of scientist/educators. For example, the journal *Molecular and Cellular Biology* (whose title means it focuses on eukaryotic cells rather then prokaryotic cells) is divided into four sections for papers on gene expression, cell growth, cell and organelle structure and assembly, and chromosome structure and dynamics. In a typical issue, sixty-five percent of the papers fill the section on gene expression.[13]

The shift in definition is also illustrated in common usage. In an announcement of an interdisciplinary seminar for Albany area faculty, a highly respected professor of biology and director of the Center for Molecular Genetics at the State University of New York at Albany used the terms *molecular biology* and *molecular genetics* interchangeably. The seminar, entitled "Molecular Biology and Evolution," was advertised as follows:

> Only in the last few years has the field of *molecular biology* made an impact on the study of evolution. Recently, several developments in *molecular biology* have made contributions of both a theoretical and technical nature to the study of evolution. . . . This seminar is for any teacher or researcher in the biological sciences who would like to explore both *molecular genetics* and evolution. (Emphasis added)[14]

The title and first statements use *molecular biology* and then, at the end, the term is converted to *molecular genetics,* suggesting that the more encompassing *molecular biology* is conflated with *molecular genetics* in the end.

The central focus of current molecular biology as represented in the textbooks cited is that DNA, the major component of genes in most organisms, is considered "the controlling molecule of life." The following statements are from the opening pages of the new edition of Darnell, Lodish, and Baltimore's *Molecular Cell Biology:*

> Successful studies of gene structure and the genetic control of protein synthesis (to be fully discussed in later chapters) form the recent history of molecular cell biology. . . . [By 1952,] a small group of informed scientists knew that DNA was the controlling molecule of life. . . . The modern era of molecular cell biology has been mainly concerned with how genes govern cell activity *and how proteins carry out specialized tasks. . . . the protein products of* certain genes regulate the activity of other genes. This principle finally unified the diverse approaches to the subject that we now call molecular cell biology. (Emphasis indicates words added to new edition as compared to original text.)[15]

The current and predominant definition of molecular biology as the study of "how genes govern cell activity" embodies several key elements that relate to feminist concerns about beliefs in the naturalness and rightness of dominant/subordinate relationships. I will address the following three issues in the remainder of this chapter:

(1) DNA is presented as the "controlling molecule of life." This raises concerns about the inherent privileging of heredity as the most important single purpose of "life" and about implications for inadvertently supporting biological determinism;

(2) The overriding representation of relationships among subcellular components is one of hierarchical control; and

(3) The given definition of a "gene" as a segment of DNA that codes for a protein reflects an assumption about the distinction between "genetic" and "nongenetic" or DNA and not-DNA analogous to the polar concept of nature versus nurture.

The Paradigm of Control by Genes

"Control" in which domination and subordination describe a fundamental and asymmetrical power relationship is a major focus of feminist critiques of society (see chapter 2). The first time the reader sees the term *control* in the Darnell, Lodish, and Baltimore textbook is on page 9, in a section on the history of the merging of genetics and biochemistry. Referring to Archibald Garrod's conclusions in 1909 that connected the inheritance of a disease, alkaptonuria, with the absence of a particular enzyme (homogentisic acid oxidase), Darnell, Lodish, and Baltimore assert: "But it was too early in the history of genetics or biochemistry for the realization that *genes do* in fact *control* the structures of enzymes" (emphasis added).[16] This is followed by a subheading highlighted in red: "*Drosophila* Studies Establish the Connection between Gene Activity and Biochemical Action; *Neurospora* Experiments Confirm That One Gene Controls One Enzyme."

Variations on the concept of a unidirectional (genes determine proteins) causal relationship between gene and protein alternate with the standard terminology of "control." In this way, genes both *control* and *determine* proteins.

> They concluded that each mutant cell that could be restored to growth by the addition of a single compound carried a defect in a single gene that impaired the production of an enzyme necessary for a single metabolic step; in other words, one *gene was responsible for* one enzyme. (Original emphasis in text)[17]

The further unfolding of the history of molecular biology stresses the significance of the feedback loop: *the regulation of genes themselves*—by proteins. The red subheading reads, "The Activity of Genes Is Highly Regulated by the Protein Products of Other Genes." The question seems to be: Genes control everything in the cell, but what controls the genes? And the section explains:

> The work of Beadle, Tatum, Lederberg, and others in the 1940s and early 1950s had made it clear that *genes encode proteins,* but the *means by which gene activity was controlled was still unclear.* Yet it was widely supposed that the differences between cells having the same genes were due to differences in the activities of those genes. Two French scientists, François Jacob and Jacques Monod, proposed that *the protein products of certain genes regulate the activities of other genes.* This principle finally unified the diverse approaches to the subject we now call molecular cell biology. The concept of the cell had evolved a long way from its original characterization as a simple unit of living matter: the cell had become an organism in which *the controlled and integrated actions of genes produce* specific sets of *proteins that build* characteristic structures and *carry out* characteristic enzymatic activities.[18]

In a conceptual system that elevates genes above other components and generally assumes hierarchical relationships, "genes which regulate the activities of other genes" are more important than other genes. And it is the principle of *genes (through their "products") regulating other genes*—rather than cellular proteins and nutrients or other molecules regulating gene activity through communal feedback loops—that the authors credit with unifying the study of bacterial genetics with eukaryotic genetics to create the new term, "molecular cell biology."

The language of regulation involving cell components other than DNA contrasts with the language of genetic control:

> A number of weaker noncovalent chemical bonds and interactions *help to determine* the shape of many large biological molecules and to stabilize complexes composed of different molecules. . . . Enzymes are biological catalysts that *generally facilitate* only one of many possible transformations that a molecule can undergo.
>
> . . . a protein *involved in controlling* gene expression . . .
>
> The activity of CAP is *regulated by* a small molecule, cyclic AMP (cAMP). . . . Each bound cAMP contacts both subunits, and binding is believed to *influence* their relative orientation. This change in quaternary structure is *probably responsible for determining* whether the protein can bind specifically to DNA. . . . Antibodies provide another excellent example of the way *distinct functions are performed* by distinct domains in a molecule . . .[19]

Occasionally, the stronger language of control *is* applied to components other than the gene: "Many protein functions are *controlled by* reversible side-chain modifications."[20]

These examples illustrate that regulation and control occur in many different and significant ways in the cell at the molecular level of organization and activity, and that the discourse of molecular biology does acknowledge this variety. The opening of a chapter that describes the structure of macromolecules identifies proteins as exerting significant control in the cell: "The chemical reactions that constitute life processes are *directed and controlled by proteins.*"[21] So the text tells us that proteins control "life processes," but it has already established that genes control proteins. Clearly, in this framework, genes are at the pinnacle of the complex process of regulation of life in cells. Here, while the authors represent life processes as consisting of chemical reactions, a view closer to that of traditional biochemistry, the language used in the textbook sets up a hierarchy of what is really important in "life."

While the language of regulation and interaction among macromolecules and subcellular components sometimes reflects a reciprocal relationship, the centrality of the gene and its greater importance as "the controlling molecule of life" proposes a paradigmatic hierarchy within molecular biology. It is the pattern of consistent use of the language of genes actively controlling proteins (as compared to the only occasional use of similar language for proteins and rare use for other components) that indicates the place of genes at the top of the hierarchy of control of the cell.[22] Active control by genes is contrasted with a more passive role of the components, usually proteins, being controlled. That active/passive difference is what distinguishes the power relations in a dominant/subordinate relationship.

Genes Are at the Top of the Hierarchy

Darnell, Lodish, and Baltimore envision genes at the top of a hierarchy of active control, while proteins do the menial work:

[G]enes do in fact control the structures of enzymes. . . .

[G]enes produce [while] proteins build . . . structures and carry out . . . activities. . . .

Molecular biologists have used the exciting new techniques of molecular genetics to isolate hundreds of the *genes responsible for many important elements* of cell structure and function. . . .

Proteins are seen to *serve as the key working molecules* of biological systems and *nucleic acids to encode information* for protein synthesis. . . . *[N]ucleic acids, the substances that preserve and transmit genetic information . . . proteins, the products* generated from the transmitted information. . . . *Proteins are the working molecules of the cell.* They catalyze an extraordinary range of chemical reactions, provide structural rigidity, control membrane permeability, regulate the concentrations of metabolites, recognize and noncovalently bind other biomolecules, cause motion, and control gene function.[23]

The laboring proteins that keep things going in cells are less important than DNA's important work of guarding and giving directions to the cell from the nucleus. The difference in jobs and status conjures up a worker/manager relationship imbued with values about class differences. Indeed, the factory and production metaphor is common in the language describing protein synthesis, with ribosomes as the machinery, messenger RNA as the copy of the original blueprint from DNA, and the product as specific proteins made to specification.[24] In an article for *Science 84,* a popular science journal originating from AAAS as a more accessible version of *Scientific American,* David Baltimore turned to the factory as a guiding metaphor for the fundamentals of molecular biology. He credited physiology with the powerfully successful model of organisms as factories, contrasting the physiological model with that of genetics and then bringing them together:

The approach of genetics, on the other hand, is to ask about blueprints, not machines; about decisions, not mechanics; about information and history. In the factory analogy, genetics leaves the greasy machines and goes to the executive suite, where it analyzes the planners, the decision makers, the computers, the historic records. It is the business school approach rather than the engineering approach. . . . But there was no understanding of the link between the executive suite and the factory floor. Watson and Crick provided that link.[25]

The title of the article identifies "The Brain of a Cell" as the genes on chromosomes, and then more specifically, as the history of this science is unfolded, the DNA:

Once the structure of DNA was found, the link between physiology and genetics was soon made. The cell's brain had been found to be a tape reader scanning an array of information encoded as a linear sequence, that is ultimately translated into three-dimensional proteins.[26]

Indeed, the physiological model of the factory leads us back to Martin's analysis of themes of production, of centralized control, and of hierarchy, with the brain at the top.

More subtly, the authors' language grants DNA a higher position relative to protein in the statement: "DNA was the controlling molecule of life," using "DNA" as a universal singular. (DNA is actually DNAs or, more precisely, segments of DNA.) In contrast, assertions like "proteins build," "proteins are seen," and "proteins are the working molecules of the cell" establish proteins as common, general, and nonspecific. The language subtly grants DNA a universal status as "the molecule," even though DNA molecules are neither singular nor uniform, while casting proteins as plural and generic. Ultimately, a variety of proteins is assumed, while "the DNA molecule" embodies an abstract universalized concept.

A Hierarchy Exists among Genes as Well

With a hierarchy of subcellular components and macromolecules entrenched in molecular biology, scientists further superimpose hierarchical control onto genes themselves. In a heading in their chapter on the molecular biology of development, Watson et al. highlight a ranked order among genes, with "A Hierarchy of Genes Controls Development":

> The goal of molecular biology studying development is therefore to discover which genes control the expression of other genes.
>
> Traditionally, one of the problems that has confounded and confused the molecular study of development is that cells usually become "committed" or "determined" to differentiate long before any actual morphological differentiation is apparent. . . . Once we realize that genes are arranged in a hierarchy, it becomes clear that in most cases, *commitment or determination* corresponds to the expression of a controlling gene, while differentiation reflects the myriad molecular consequences of that initial developmental decision.[27] (Emphasis in text.)

This statement makes explicit what has been implied: the extension of control over more components and processes confers a higher value. The controlling gene exerts a primary influence by determining the course of development, while the other genes involved in the "myriad molecular consequences" that constitute the actual process of differentiation are clearly secondary.

While most scientists no longer use terms such as "master" and "slave" to describe DNA's position of power, they have dubbed the genes that code for enzymes running the cell's metabolism "housekeeping genes."[28] This term clearly implies a routinized and less significant kind of gene, its value analogous to that of women's wifely and motherly roles, which are not credited in the nation's economy or deserving of compensation in social security benefits, higher education, or hiring preference (as armed forces veterans receive). Maintenance is less important, and scientists use the gendered metaphor "housekeeping" to capture that status.

Genetic Versus Nongenetic: Another False Nature Versus Nurture Dichotomy

Conceptual and operational definitions of a gene differ according to the focus of the subfields of biology. A gene is variously a unit of inheritance or a unit of selection. Molecular biologists define a gene as a unit of transcription, a sequence (or sequences) of DNA that codes for a protein or RNA—plus any DNA sequences involved in gene expression, such as the promoter or operator regions that are "controlling sequences" of gene expression.[29] Molecular biology asserts that a particular DNA sequence determines or codes for a corresponding messenger RNA sequence. The RNA sequence then directs the translation of an amino acid sequence that makes up a particular protein. This amino acid sequence is linear and is called a polypeptide, the primary structure of the protein. One of more polypeptides then fold in a particular way, depending on their interaction with the local environment. The folding thus "determines" (more accurately, influences) the structure and function of the protein. In a similar way, when a piece of DNA codes for a functional RNA molecule that is not a "message" for translation into a protein, that DNA sequence determines a corresponding sequence of ribosomal or transfer RNA, RNA molecules that fold into a specific structure. Hence, a gene is that piece of DNA that codes for a particular protein or RNA.

One of the assumptions built into the paradigm of DNA as "the controlling molecule of life" and "the genetic material," then, is that a "gene" is a segment or segments of DNA that code(s) for a protein.[30] The important DNA is distinguished from the "junk": "Amidst this sea of unstable, probably nonfunctional DNA lie the genes, islands of coding DNA recognized by classical genetics." More precisely, "The concept of the gene as a biological entity—that is, a heritable function detected by observing the effect of a mutation—is still valid. However, according to our current molecular definition, a gene consists of all the DNA sequences necessary to produce a single peptide or RNA product."[31]

In contrast to these accepted characterizations, the structure and function of genes would be more accurately represented by a definition that includes a segment of DNA with the proteins (and carbohydrate, phosphate, and other moieties as well) complexed with it, as well as the processes involving other molecules such as enzymes and cofactors. These molecules interact with the nucleoprotein complex, while it functions as a repository of some of the information, the linear code, that shapes the production of the corresponding RNA and proteins. Support for this alternative view comes from the full-page picture [see figure 6-1] that faces the introduction to the history of molecular biology in the Darnell, Lodish, and Baltimore textbook: it is labeled "λ-Repressor-DNA complex," a complex which is, in my view, a particular gene in the process of functioning. This definition of a "gene" would counter distortions arising from reductionist reification of the complex dynamics that constitute "life" at the molecular level.

In the traditional view, a narrow, control-dominated picture of the gene enforces the conceptual separation of genetic from nongenetic, decontextualizing the gene with an exaggerated distinction that reflects the narrowing definition of "genetics." Yet the boundaries between the DNA sequence and its environs, including molecules that interact with it and influence its functioning, are not distinct. To define "genetic" as the sequence of DNA bases— even though gene expression is described in conjunction with the other kinds of molecular activities associated with the functioning of genes—strips DNA of its actual context: in dynamic relationship with small and large molecules and ions in a cell that in turn is affected by the environment and activities around it, including interactions with adjacent cells.

The genetic/nongenetic dichotomy, presented as a clean either/or relationship, is as arbitrary as the analogous view of biology as a fixed entity that can be separated from that with which it is contiguous. Just as "nature" is not actually separate from "culture," or "nature" from "man," or "nature" from "nurture" or environment, the separation of "genetic" from "nongenetic" depends on an arbitrarily chosen boundary. Our cultural nature/nurture split reinforces an ideology of separation and alienation of an individual from its environment as well as from other organisms. Yet organisms do not exist apart from their environs or from other organisms. Not only are organisms surrounded by and embedded in a dynamic interaction with their environs— and in that sense are contiguous with it—but we are contiguous with the environment from the inside as well, whether through our digestive and respiratory tracts, our skin pores, or the network of endoplasmic reticulum throughout the cytoplasm of many types of cells. A human body is many organisms, most of which are necessary for a healthy life: *E. coli* in the lower intestine, microorganisms on the skin. These flora and fauna produce materials useful to our bodies and help maintain conditions, such as pH and water regulation, which prevent disease-causing infections.

A very different psychology of self and other would understand our beings as open to and connected with the environment around us through our external and internal surfaces, as well as what we project of ourselves (our exhalations, body heat radiation, wastes, etc.). Further, a historical dimension to this dynamic would be added if we considered the "recycling" of atoms and energy through living and nonliving things. This interactive and contextual conceptualization of living beings (quite accurate scientifically) encompasses consideration of the effects of our "life" on the rest of the planet's "life,"

Figure 6-1. A gene as a complex of DNA and proteins. "λ-Repressor-DNA complex," from *Molecular Cell Biology 2/E,* by Darnell, Lodish, and Baltimore, facing page 1. Copyright © 1990 by Scientific American Books, Inc. Used with permission of W. H. Freeman and Company.

a view that would have us addressing and perhaps preempting the planet's environmental crises much sooner and more effectively.

The dominant view of molecular biology reinforces at yet another level of organization an arbitrary, dualistic ideology that separates "genetic" from "nongenetic" and at the same time reduces "genetic" to a context-stripped piece of DNA. As scientists reified and reduced "nature" in the genetic sense to the DNA sequence, then "nurture" or the environment surrounding the gene became everything but the DNA. The decontextualized and reductionist definition of the gene both reflects and recreates the nature/nurture fallacy at the molecular level of organization in cells. And the high status of this field legitimizes this powerful, though false, dichotomy.

Is This Representation Distorted?

Consider a hypothetical approach defining "life" as the functioning of proteins, instead of DNA. As the textbook statement, "The chemical reactions that constitute life processes are directed and controlled by proteins,"[32] suggests, it is not incorrect to say that proteins control life in their ability to function as enzymes, in their contribution to enzyme action by the physical structures constituted by them and other components, holding things in place, and in their control over gene expression. Why not, then, organize molecular biology around the view that proteins control everything in the cell? Why not see molecular biology as the study of how proteins order life processes, the chemical reactions of the cell? Or, if we revered the power of smallness, we could organize molecular biology around the tiny magnesium ions ($Mg++$) that are essential for replication of DNA and other activities of the cell. These statements are as accurate as those made about DNA, but they derive from a different focus and a different set of assumptions and values about what constitutes "life," while still singling out one component as most important.

Instead of focusing on command and control, molecular biology could focus on metabolism, the maintenance of physicochemical beings, or the conversion of energy from sunlight through plants to animals, making the central subjects of molecular biology chloroplasts and mitochondria and other energy-conversion systems in living beings. Like the dominant perspective on the primacy of DNA, none of the statements is scientifically inaccurate in the strict sense of not correlating with the information we have about what is happening inside cells, but they are *partial* and thus *misleading*. Here, then, is an example of the *social* construction of science; that is, we can recognize that we have selected only one of several possible ways of organizing our knowledge about the cell and molecules—but the choice by leaders in the field is not random. The selection of a construction that represents DNA as the top of a hierarchy of control over life is based in a cultural paradigm that does not, ultimately, benefit either science or society.

Alternate views of "life"—metabolism, energy-conversion, autopoiesis (self-maintenance with or without reproduction),[33] and those that value complex interdependence—help us appreciate that the view that defines "life" solely in terms of genes and heredity is misleading. I suggest that a more *accurate* paradigm would value the functioning of all components of the cell, since they work together to harness energy ultimately from the sun to maintain complex structures, to metabolize, sometimes to reproduce, to respond to the environment, in short, to create the many characteristics we call "life."

In the current view of molecular biology, "gene expression" constitutes the way life functions, while the gene, whose linear bases are presented as the ultimate code for life, *is* life. The dominant framework gives the powerful concept of "information" a higher value than other functions in the cell and organism.[34] From a different perspective, however, the gene could be understood as the storage place for information, just as glycogen or starch is the repository of one form of energy for the cell.

Instead, molecular biologists claim that the information in the sequence of the bases in a piece of DNA confers *specificity* by directing the synthesis of a unique protein, and it is this specificity of DNA that confers the special status of DNA.[35] But why does one piece of DNA, one gene, "tell" us more than an enzyme, a coenzyme, or a magnesium ion essential as a cofactor in a very specific process of binding? If we place DNA into its context in the cell and shift the focus away from DNA itself, we find that specificity arises in protein synthesis not only from the base sequence of DNA, but also by the way that DNA is copied into a messenger RNA molecule, interactions with specific transfer RNA molecules, interactions with specific parts of ribosomes, interactions with specific enzymes and cofactors, all within a microenvironment of a certain pH and water concentration. Ultimately, it is more accurate to say that proteins are produced from the *interacting specificities* of transfer RNA molecules, messenger RNA, ribosomes, DNA, enzymes, cofactors, energy, all affected by the "local" conditions that are, in turn, affected by the contiguous macroenvironment. At the molecular level, as at other levels of organization, the genotype does not produce the phenotype in a direct, one-to-one correlation, yet this standard biological principle is dismissed.

Similarly, scientists frequently use the term "gene product" as a shorthand for the RNA or polypeptide "copied" and "translated" from DNA. The relationship of such "products" to the genes that "code for" them is actually a consequence of the evolutionary history in complex (cumulative, nonadditive) interaction with the specific microenvironment at each moment in the life of each ancestral organism.

With this broader, more inclusive concept of "product," the mind's picture of the cell can change the emphasis on the genes and a (usually) one-way direction of information flow from genes—to proteins—without losing an appreciation for the part that DNA plays. It is in this sense that scientists such as Ruth Hubbard assert that the pattern of privileging DNA in a centralized hierarchy of life is scientifically incorrect, because "in a complex system of reactions (such as protein synthesis) which requires many components and conditions that interact in nonadditive ways and that often are interdependent, it is wrong to single out any one substance or event as causal to any other."[36]

In a nonhierarchical reconstruction, genes are only one part of a multifaceted, multidirectional hologram of "life." The term *gene product,* as currently used, narrows the concept to focus on the importance of DNA above the other components and processes of the cell and organism. The result of this narrow scope of "life" is a misrepresentation, a one-dimensional (that is, information transfer) cause-and-effect process that leaves out biological history and the context in which DNA exists and operates. The terms *gene, gene product,* and *gene expression*—because they are used within a conceptual framework of the current biology—increasingly become crystallized into "things," rather than fleshed out as complex processes.

If we say that it is equally or even more accurate to characterize the life of the cell in a nonhierarchical, inclusive, multiplistic framework that does not set one component over others as primary cause, then on what basis should we choose our paradigms? In Keller's analysis of slime-mold differentiation, she argued that both models (the hierarchical, centralized control assuming inherent difference versus steady-state equilibrium in an environmental context) were equally valid, yet scientists chose the pacemaker concept as the predominant working model. Martin's analysis of women's physiology recognized centralized control of the menstrual cycle by the brain as the predominating framework, although alternative characterizations of the relationships among the tissues, hormones, and organs are equally correct and are used in non-Western representations of physiology. The organization of components in complex processes, the meaning of control and regulation, and the value placed on different components are all skewed in one particular direction in the study of living beings. This is not an accidental distortion, not a random choice of models. Rather, it is a repeated pattern of preferred rules—the commitment to centralized control, to hierarchical organization, to difference as dominant-and-subordinate.

What might appear to be just one "choice" for a workable paradigm or metaphor is actually part of a systematically misleading distortion that constrains the way that scientists conceptualize the subject matter and skews the direction of research. In a nutshell, only if the gene is the sole determinant of life does it become possible to look to genetics alone for solutions to problems such as illness, food shortages, and human imperfections, as the Human Genome Project promises. On the other hand, if the determinants of what constitutes and directs life are presented as a balance among metabolism, energy conversion, and reproduction, in dynamic interaction over time with the environment in which life occurs and of which it is a part, then the search for solutions becomes similarly multifocal, stressing the environmental context as much as the internal environment, and requiring consideration of complex interactions and transformations.

From the point of view of science—its accuracy in representing nature and its directions of development—prior commitments to limited and repeated themes promote partial and distorted knowledge and narrow research agendas. The distortions that result from ideological bias in biology feed back into society to reinforce a view of "difference" as inherently asymmetrical and naturally hierarchical, with consequences for subordinated groups. From the point of view of the struggle against societal inequities based on race, class, and gender, we must systematically reassess recurrent conceptual patterns, and revise our paradigms.

DNA-Dominated World

The paradigms that assume life is controlled and directed by DNA and that molecular biology equals molecular genetics are so pervasive within scientific literature that they give a particular weight to the outmoded dogma of molecular biology: DNA to RNA to protein. A mini-review in *Cell* brings to the fore the assumption of the domination of DNA over life.

Harold Varmus's mini-review concerns current research that identifies reverse transcriptase activity in different bacteria, adding prokaryotic organisms to the growing list of living beings with reverse transcriptase activity.[37] Varmus places recent developments within the historical and conceptual context provided by Howard Temin's "contentious idea" of twenty years earlier, that DNA was transcribed from RNA templates in RNA tumor viruses—in short, in the "opposite" direction from the Central Dogma of DNA to RNA to protein, which is what earned it the name "reverse" transcriptase.[38]

While considering the significance of reverse transcription in bacteria, Varmus makes some loaded statements: "Indeed, since it is now fashionable to assume that the first genes *were in the form of* RNA, it may be necessary to posit reverse transcription as a crucial early step in the evolution of *our DNA-dominated world* (reviewed in Varmus and Brown, 1989)" (emphasis added).[39]

The use of the phrase "it is now fashionable to assume" adds a bit of spin, implying that the significance of RNA is more fashion than fact and hinting that, while RNAs may appear increasingly to be significant in the origins and evolution of life on this planet, DNA remains preeminent. A simple turn of phrase, then, reinforces the assumption that RNA should never be considered as a dominant form: "that the first genes were in the form of RNA." Why not say "were RNA," language expressing an exact equivalent of gene with RNA, rather than the qualifier "in the form of," which implies that RNA is not the real thing? Why "in the form of RNA," if not to protect the prerogative of DNA as the master molecule?

What does it mean to say that biologically we live in a "DNA-dominated world"? In the strictest sense, Varmus may be stating that the vast majority of living beings on earth have DNA as their genetic repository (many viruses have RNA as theirs). But why use the term "domination," then, when "ubiquitous," a word Varmus uses elsewhere in the text, is more accurate and appropriate? In an investigation of the impact of gender and related belief systems in the sciences, concerns about the pervasiveness of a worldview that accepts as natural and justified the domination of one superior type over another inferior type make it necessary to question this representation of "our world" as DNA-dominated. How does this description function within molecular biology? How does it function in our society, with its long and destructive history of insistence on superior/inferior pairs, and its use of biological determinism, among other concepts, to support the domination of one group over others?

In the last part of the review, Varmus delineates several significant issues he feels are raised by the new evidence about reverse transcription. The three significant issues he lists first are these: the nature of genetic elements in relation to evolution; the functions of msDNA (multicopy, single-stranded DNA) *in relation to genes* (rather than, for example, structure/function insights or interrelations with other functions and structures in cells); and other *genes* that might be involved in this complex operation. Placed *after* the exploration of these three issues, and with less space devoted to them—are nongenetic aspects of the topic: biochemical properties of the macromolecules and the functional interdependence of the discovered nucleic acid-protein complexes.[40] This example eloquently supports my major assertion: that DNA, as the gene and the most significant aspect of molecular life, remains the organizing character of the new molecular cell biology, even while scientists elucidate the more interdependent aspects of molecular life and research.

I think it is important to recognize that complementary methods and approaches *are* increasingly valued and that what might be analogous to an "interdisciplinary" approach to research is emerging, at least in the articles published in *Cell*.[41] However, in spite of this move away from single causes and single research approaches, the underlying devotion to DNA and the gene remains.

It appears that genetics (and evolution) is so deeply assumed as the way of thinking about the meaning of life biologically that it has simply become a given for many researchers and educators in this field. To break out of these constraints, scientists can become aware of this assumption that has become more and more invisible and can reexamine it with the healthy skepticism that fuels much of scientific work—to do so in grant proposals, journal articles—especially reviews—and in teaching. With this awareness and questioning brought into classrooms and with students engaging in this re-visioning of molecular cell biology from the first, we can take advantage of the fresh perspectives of the uninitiated to explore alternative and generative views. These critical views represent perspectives with which students and educators/researchers should be acquainted to be well-educated in the life sciences now and in the future.

Many scientists would probably agree that the most creative ways of thinking in our fields have included seeking to enlarge our views of what we study, taking as much information and as many possibilities into account as we can. Such an approach overcomes the parochialism often produced by the necessarily intensive study of one small piece of a larger picture. I suggest that feminist perspectives offer another impetus to the goal of enlarging our understanding of nature.

In support of an openness to a range of perspectives in molecular biology, I offer the concluding paragraph of Varmus's review, indirect though it may be. Varmus, appointed director of the National Institutes of Health, stresses a readiness to look forward to surprises and idiosyncrasies in our understanding of the molecular world:

> The idiosyncrasies of bacterial reverse transcriptase enlarge our notions of the forms DNA synthesis can assume now, and might have taken in the past. Yet other strategies are reflected in the recent report that a telomerase [an enzyme] adds repeats of d(TTGGGG) [a specific segment of DNA] to Tetrahymena [a single-celled protozoan] chromosomes using a small, enzyme-associated RNA as template (Geiger and Blackburn, 1989). Such a variety creates great expectations for further surprises. Will we soon be reading about retrophages? About enzymes that can synthesize both RNA and DNA from RNA templates? Stay tuned.[42]

Some of the biggest "surprises" that can be discovered by removing the blinders of dominant paradigms have been overlooked or ignored by scientists: "sex" is not the biologically fixed bipolar entity that most of science, like much of society, takes it to be; and cultural beliefs about gender pervade and distort our scientific understanding of the meaning of life at the cellular and molecular levels.

In some cases, scientific metaphors or paradigms are "wrong" in a technical sense—for example, when bacteria are labeled "male" and "female," or when "sex" is superimposed onto macromolecules, such as hormones or carbohydrate metabolites. In other cases, such as a centralized hierarchy of control of life at the molecular and cellular level, it is the power of the metaphor that systematically skews conceptualization of the cell and the meaning of living processes. The next chapter illustrates that molecular biologists have chosen deeply embedded assumptions as organizing principles for molecular biology, setting priorities for research in areas such as the understanding and curing of cancer.

7

Power at a Price

UNIFYING PRINCIPLES

A group of techniques . . . molecular genetics . . .
serve[s] to unify all experimental biology in language
and concerns.
> —James Darnell, Harvey Lodish, and
> David Baltimore, 1990[1]

Physicists and theoretical chemists, who calculate energy levels in atoms and molecules, do not have access
to more fundamental truths than have molecular biologists, who study the structure and sequence of genes
on the chromosomes. Nor are the descriptions molecular biologists provide more fundamental than those of
biologists who study cells or organisms. Biologists do
not probe deeper realities than anthropologists and historians, just different ones.
> —Ruth Hubbard, 1990[2]

Perhaps we're only "genetics mechanics" today.
> —Sidney Brenner, 1983[3]

Foundation for a "New Biology," Proposed in *Molecular Cell Biology*

James Darnell, Harvey Lodish, and David Baltimore claim that a set of techniques called recombinant DNA technologies, rather than a concept, should function to unify all of experimental biology. This explicit proposal of a unifying principle for all of molecular biology has significant ramifications for all of biology's subfields. In the first edition of *Molecular Cell Biology,* the authors announced the foundation of a "new biology"[4] in which three previously distinct fields were to be reorganized and subsumed under the new molecular cell biology. The 1990 edition repeats and emphasizes this point that what is "required" is "a reformulation of a body of related information formerly classified under the separate headings of genetics, biochemistry, and cell biology." This reformulation of these three disciplines does not occur around a unifying concept or theory but from a "group of techniques collectively referred to as molecular genetics." The reasons for giving molecular genetics this central role in reorganizing biology are the "powerful analytic force" of the techniques and their ability "to unify all experimental biology in language and concerns." The second reason is actually circular; by placing the techniques at the center of biology, the language and concerns of molecular genetics become the filter through which nature will be studied and conceptualized. Indeed, the authors state, "With the tools of molecular genetics, genes for all types of proteins—enzymes, structural proteins, regulatory proteins—can be purified, sequenced, changed at will, reintroduced into individual cells of all kinds (even into germ lines of organisms) and expressed there as proteins." And, in circular justification, they again assert that "[m]ost of experimental biology now relies heavily on molecular genetics."[5]

What is particularly significant about their declaration is that, while the rationale for reformulating those areas of biology appears to be the new information obtained in the past decade (particularly that about eukaryotic cells) with recombinant DNA techniques,[6] it soon becomes clear that the analytical power of recombinant DNA techniques, not the new information, is *itself* seen as sufficient reason for the proposed reorganization of biological knowledge into a new hierarchy. Circular reasoning starts with the power of this particular technique to provide certain kinds of information and ends up using the set of techniques for organizing knowledge systems at the cellular and molecular levels.[7]

In case there was any doubt about the new dominance of recombinant DNA techniques, the authors open the preface to the second edition with the following point:

> We asserted in the preface to the first edition of this book that *the reductionist approach and the new techniques of molecular biology would soon unify all experimental biology.* Now, four years later, perhaps the only surprise is the speed and completeness with which biologists from fields formerly considered distant have embraced

the new experimental approaches. . . . Our contention four years ago, reaffirmed now, is that *the teaching of biology must reflect a unified experimental approach.* The education of biology students should be shaped by it from the beginning.[8] (Emphasis added.)

In *Molecular Cell Biology,* the authors have articulated and formalized what has been taking place within the field since the 1970s. While other textbooks may not be as explicit in calling for a reorganization of knowledge in biology, most of them proceed from the same underlying assumptions about the power and significance of recombinant DNA technology.[9]

Molecular genetics techniques are not the only ones credited with great advances in molecular biology, but they are the only ones elevated to the status of a central technology for reorganizing the study of life. The paragraph following the declaration points to "comparable advances" in culturing cells and in "sophisticated instrumentation," such as electron microscopy and computers. The chapter devoted to the "tools of molecular cell biology," "molecular technology," or, in the second edition, "manipulating macromolecules"[10] makes the following statement about the importance of radioactive tracers: "Almost all experimental biology depends on the use of radioactive compounds."[11] Yet only one set of techniques is singled out as an organizing principle for the field of biology, in part due to the view that "many deep biological secrets were locked up in the sequence of the bases in DNA."

> Almost overnight, this group of techniques, often collectively called *molecular genetics,* became the dominant approach to the study of many basic biological questions, including how gene expression is regulated in eukaryotic cells and how protein or domains of proteins function. The power and success of the new technology have raised high hopes that the practical use of our ever-increasing biological knowledge will bring many benefits to mankind.[12]

Why is it so important to claim *one* dominant approach? Although each subfield of biology declares a central place for itself relative to the other subfields, these authors want molecular genetics to be the *sole* unifying principle for studying life. Within this framework of the preeminence and power of recombinant DNA techniques, reference to "older procedures still widely used in molecular experiments today" sounds almost patronizing.

There is little doubt that each historical epoch in the development of this science has advanced, in part, through the new opportunities brought about by major new techniques such as enzyme assays, spectroscopy, radioactive tracers, scintillation counting, and nucleic acid hybridization. Recalling the importance of specific techniques in the descriptions of various subfields,[13] what justifies singling out the techniques of recombinant DNA as the only ones worthy of the label *technology?* Such categorization clearly implies the elevated importance of those techniques over other equally necessary and effective ones. I suggest that raising recombinant DNA technology to the status of an organizing principle for a new molecular biology and then a new biology (no, *the* new biology) closes down other ways of approaching the study of life at the cellular and molecular levels. While that would not have to be the case, my analysis below suggests that a *singular* approach is being promoted at this time. Furthermore, the oddity of using a technique, rather than a concept, as an explicit unifying principle raises fundamental problems about the epistemology of this stance (is there only one way of knowing molecular biology?) and suggests hidden contradictions about the meaning of molecular biology that must be investigated.

Evelyn Fox Keller has argued provocatively that the genetic origins of molecular biology came from scientists (mainly physicists) committed to a reductionist view of life intertwined with the desire to control life by solving its mysteries ("deep biological secrets were locked up in the sequence of the bases in DNA"). Life's secret, cast as reproduction, is held by women; dominance relations in society compel Western male scientists to ferret out the last vestiges of female power. In this light, it is not surprising that the techniques of molecular genetics are the chosen organizing principle for the definition and study of "life."[14] Other views that emphasize the close relationship of science and capitalism would also place the new molecular biology into a context of social control.[15]

The pronouncement made by the authors of *Molecular Cell Biology* raises to a new level of concern several of the problems cited elsewhere in this book. These problems include presenting science as a nonpolitical and value-neutral endeavor based on objective methods as a means to accurate knowledge, with techniques in particular seen as value-neutral; privileging molecular genetics as the lens for understanding and explaining "life," with the attendant problems of biology as a rigid determinant of life's activities and behavior, along with a polarized view of nature in relation to nurture; and consequently narrowing the frame of reference for the meaning of "life."

Techniques Are Seen As Value-Neutral. By defining molecular genetics as a set of techniques (that turn out to be primarily recombinant DNA technologies), rather than an explicit conceptual framework, the authors reinforce the traditional belief in scientific objectivity and the neutrality of science. Although advances in techniques have had a central impact on the development of all the fields in biology (the light microscope, for example), it seems odd that a whole field is explicitly organized around a set of techniques, rather than a concept. For example, advances in microscopy have contributed enormously to our understanding of cells and the organelles and substructures within them, but microscopy is not claimed as a unifying principle of cell biology. Instead, structure/function relationships of subcellular structures and a "cell theory" that holds that the cell is the basic structural unit of "life" are explicit principles of the field of cell biology.

What does the dominance of a technology mean with regard to hidden values and beliefs? Techniques seem to be value-neutral, but in fact they are big business in both economics and epistemology. Any technique or technology is designed to provide specific kinds of information and produces certain material consequences from its methodology. Corlann Bush proposes that tools and technologies have valences, "analogous to that of atoms that have lost or gained electrons through ionization. A particular technological system, even an individual tool, has a tendency to interact in similar situations in identifiable and predictable ways."[16]

Recombinant DNA technologies are valenced toward seeking answers in the anatomy of DNA molecules. Ideological definitions and hierarchies give particular meaning to the anatomy of DNA molecules; DNA is a gene, and "The Gene" controls life.[17] The techniques of molecular genetics provide certain information about "the gene" and its "expression" in a scientific context that narrows the meanings of those terms, reinforcing artificial separations of genetic and nongenetic, while distorting the relationship of genotype to phenotype, and, consequently, reinforcing biological determinism.

By accepting this formulation without question, molecular biologists sidestep the political and social implications of the principle actually underpinning the new biology; since knowledge of "the gene" and its "expression" is knowledge of "life," and since the gene is the fundamental unit of life, then the major questions in the study of biology are to be organized around the gene. With studies of the functioning of proteins that "work together to make a living cell," conceived within a framework of proteins as "the ultimate gene products," research problems are cast primarily in terms of genes and their products. Less attention is given to the other "factors," such as the organism's environment or the organism's history, which in this view can only "affect" predetermined biological processes. Ultimately, the paradigm of molecular genetics as the basis of the study of biology forces us to view all of life, including behavior and social structures, as "gene products." When you study to be a molecular biologist, very little of your education is usually spent on the ecology and physiology of whole organisms.

Not only does recombinant DNA technology set the conceptual framework of "life," it also organizes scientific research. In the recent controversy about the Human Genome Project, Bernard D. Davis and colleagues from the Department of Microbiology and Molecular Genetics at Harvard Medical School argued that the Human Genome Project, rather than expanding the sources of funds for research, has instead cut small grants for untargeted research in the biomedical sciences. Although those small projects are considered to have the greatest potential for promising discoveries, funding has decreased dramatically in the past few years.[18] Tracing the history of the Human Genome Project, Davis asserts that the idea came from an administrator in the government who was "convinced that the powerful tools of molecular biology made it appropriate to introduce centrally administered 'big science' into biomedical research."[19] Considering the conceptual and psychological power of certain models of what is naturally "best," it may not be accidental that the power of recombinant DNA techniques seems valenced toward a hierarchical, centralized organization of scientific research.[20]

Using a set of techniques as an ostensibly value-neutral approach to the study of life masks deeply political and powerful commitments, some of which are clearly implicated in the highly charged struggles around the Human Genome Project, its scientific value, fund-raising strategies, and safety and ethical issues of recombinant DNA experiments and applications. Yet the debates about potential biohazards and applications of recombinant DNA technology (which have been at times highly acrimonious) are absent from Darnell, Lodish, and Baltimore's 1,000+ pages, making molecular biology look value-neutral and free of politics. Since overt political analysis is rarely included in science courses, this depoliticization is very effective in socializing scientists to ignore or dismiss the political content of scientific work.

As Molecular Genetics Becomes Molecular Biology, the Values and Ideology of Biological Determinism Become Entrenched, but Less Visible. The current formalizing step articulated by Darnell, Lodish, and Baltimore strongly supports my earlier argument that molecular genetics as a successful methodology has become molecular genetics as an encompassing worldview of biology. The powerful techniques of molecular genetics, now applicable to animal and plant cells, are being used *not just as techniques, but as organizing principles for all of biology.*

Much of the molecular biology represented in scientific journals supports the hegemonic influence of molecular genetics within biology. We see a prime example in a new journal, *The Plant Cell,* published by the American Society of Plant Physiologists. In the table of contents of the first issue, published in January 1989, thirteen of the fifteen research articles feature "genes" or "DNA" or "mutant" in the title. Clearly, this new journal is devoted to studying plant cells primarily through the lens of genes and genetic control.[21] That scientists treat the "new biology" as a fait accompli in some quarters is evident in a new textbook: *Biotechnology: The Science and the Business,* in which the first subheading under the major section, "Underlying Technologies and Economics," is: "The 'New Biology.' "[22]

With this worldview of "the new biology" come undetected ideologies embedded in molecular genetics, intertwined with gender beliefs. The focus on nuclear or chromosomal genes as the most important units of life follows from and, in turn, reinforces a reductionist and biological determinist belief that what is coded into the DNA sequence determines what is scientifically important about the cell's activities and, hence, life. Watson's reference to the human genome as the Book of Life is further clarified by his assertion that the objective of the Human Genome Project is "to find out what being human is."[23] This fundamentalist ideology embraces a rigidity in the meaning given to "biology," since the DNA sequence must be a conservative one, both for the cell to function properly (as a product of evolutionary adaptation to its environment) and for the next generation of cells and organisms to function similarly (heredity). The ideology is also hereditarian in the belief that what is fixed in our biology is inherited, passed on unchanged to subsequent generations. "Biology" is understood as fixed and unchanging, when it is actually elastic and variable. Despite such phenomena as jumping genes, mutations and other changes in base sequence, diverse modifications in chemical constituents that affect the functioning of DNA, and associations in complexes, the belief in a rigid biology applies in the realm of DNA and genes. Some of these molecular events are part of normal processes of development and aging, while others may lead to diseases such as cancers, but all are examples of changes that DNA and genes undergo during the lifetime of an organism.

The combination of reductionism, reification, and the elevation of the stripped-down gene to the top of a hierarchy of control obscures the valid perspective that mechanisms involved in heredity and control of protein synthesis are equally dependent on enzymes, other proteins, cofactors, water molecules, structural organization, and complex interactive processes among components of the cell and the surrounding environs. A reductionist fallacy first distinguishes complex processes (protein synthesis, information transfer, DNA replication, organizational assembly) from a single original cause (such as DNA sequence and gene expression) and then conflates the complex processes with the causal entity. Reification and reductionism substitute a piece of DNA for a complex process. As a consequence of the assumptions embedded in the new biology, biological determinism, with its history of justifying sex, race, and class hierarchy and, therefore, oppression, is strengthened *even though such issues appear nowhere in an explicit form.*[24]

Cultural beliefs embedded in molecular genetics remain invisible in the "new biology." Molecular genetics conceals beliefs in biological determinism and hereditarianism, as well as a nature/nurture split and natural hierarchies of power, while extending them to all of biology.

Narrowing the Frame of Reference for the Meaning of "Life": Proposed Reorganization of Subfields of Biology. One could argue that every subfield has its focus and that the orientation of molecular biology, which places the gene in the center of the field, is just one of many approaches taken in the life sciences. That argument might be valid if students in biology and molecular biology were taught several subfields equally, valuing different approaches to understanding "life" at the level of macromolecules and in relation to other levels of organization of "life." But implicit in the call for a common language and a common set of concerns is a concomitant narrowing of perspectives.

Formal descriptions of different subfields of biology foreground certain characteristics and levels of organization, while setting others as secondary, although necessary and supporting. For example, the field description for developmental biology in *Peterson's Guide* places genetics as a subdiscipline under development, while stressing the essential interrelationship of the two designated areas. What distinguishes Darnell, Lodish, and Baltimore's pronouncement is its explicit placement of molecular genetics over all other areas of biology to function as an organizing principle for conceptualizing biological life.

The authors make the aim of their textbook quite clear in the preface. After claiming that scientists have created a new biology in the last decade, the authors explain that the purpose of their book is not just to present this new information, but to reorganize the framework within which all of biology is to be comprehended fully.

> It was our purpose to teach a one-year course that integrates molecular biology with biochemistry, cell biology, and genetics, and that *applies this coherent insight to such fascinating problems* as development, immunology, and cancer. We hope that the availability of this material in a unified form will stimulate the teaching of molecular biology as an integral subject and that such integrated courses will be *offered to students as early as possible in their undergraduate education.* Only then will students be *truly able to grasp the findings of the new biology* and its relation to the *specialized areas* of cell biology, genetics, and biochemistry.[25] (Emphasis added.)

The intention is to fit the lens of molecular genetics into the glasses of under-graduate science students at the start of their education in biology, whether teachers emphasize "the gene or the cell."[26]

The authors' goal is to be *comprehensive,* as well as to merge molecular biology with the three new *sub*disciplines ("specialized areas") of cell biology, genetics, and biochemistry. Such reorganization significantly narrows students' perspectives on the study of "life" to mean only reproduction and gene control. And among those students are not only undergraduates and graduate students in the life sciences, but also medical and dental students. Furthermore, this textbook has been used for faculty seminars on molecular biology and is certainly an important resource for science writers.[27]

The paradigm of the new molecular biology excludes other frameworks for visualizing the functioning of cells at the molecular level in relation to organismic functioning. This was brought home to me in a conversation with an audience member after a presentation on this work. Recently immersed in learning molecular biology so that she could teach science teachers the fundamentals, a teacher interested in feminist perspectives expressed her discomfort. Was I denying the truth about DNA, RNA, and protein when I challenged the Central Dogma? If so, she could not imagine an alternative view. What does it mean to have only one conceptualization of molecular and cellular life—and to be unable to place the roles of DNAs, RNAs, and proteins into a context that would allow shifts of foreground and background?[28] Those of us trained in traditional biochemistry and also in cell biology, developmental biology, ecology, and classical biology have a range of ways of conceptualizing life's activities at the level of molecules, organelles, cells, organisms, populations, and biosphere participants. How much are students training to be science teachers—or research molecular biologists—required to know about the extraordinary variety of species of life and the array of their physiological, metabolic, and developmental capabilities? As scientist-educators reorganize curricula around this new version of molecular biology, upcoming generations of students lose this diversity of perspectives.

Further, the concept of "information" promotes a reification of life proc-esses into a universal causal code that is inadequate to account for the speci-ficity and diversity found among molecules and organisms. While proteins are considered important in the work they do in the cell, and different kinds of RNAs play central roles in the translation of the linear sequence of nucleo-tides in DNA to protein structure and function, the sequence of DNA termed "the gene" holds the focus of the field of molecular biology and thus suppos-edly holds the answer to life's mysteries (disease, evolution, and develop-ment). And we are to reorganize all of experimental biology around that focus.

Understanding life as a consequence of the flow of energy from the sun or as the physicochemical reactions that maintain life or the complex interac-tions of organisms with their environments, including other organisms, has been displaced in the landscape of molecular biology and in biology in gen-eral, so that life is characterized primarily as the reproduction of genetic infor-mation, even at a time when global and local ecological issues threaten the very "life" of this planet. Recent developments in the field of molecular biol-ogy, therefore, tend to cut off alternative ways of conceptualizing life proc-esses, reducing molecular genetics with a biological determinist ideology embedded with gender beliefs.

Molecular-biology-as-genetics is rarely questioned in *Science,* other than in debates among scientists about the significance and value of the Human Ge-nome Project.[29] In a notable exception, on the thirtieth anniversary of the discovery of the double-helical structure of DNA, reporter Jeffrey Fox used the "rather traditional gathering of the most successful members of the molecular biology club" as an occasion to comment on the limitations of putting all our biology eggs in one molecular genetics basket. The article contrasted Watson, "still an enthusiastic lobbyist for molecular biology, particularly genetics," against Crick, who "by contrast, has left that subject behind and set his mind to studying the brain." DNA technology may not be "adequate to the task of understanding the intricacies of the central nervous system." Using Crick's expert status, "these tools tend to furnish a 'linear' understanding of prob-lems, in the sense that the genetic code is linear. Whether they can furnish a deeper understanding of the brain—or, for that matter, of other puzzles such as differentiation and development—is another question altogether."[30]

Over one-third of the report treats the limitations of depending on the tools of molecular genetics, noting that such shortcomings are "seldom considered any more because the molecular biologists have enjoyed so many successes" and because "the rush to popularize the appropriate techniques and cash in on them tends to obscure the inscrutability of the information often thereby produced." Raising the specter of the contrast between true, all-seeing knowledge ("all there is to know") and insignificant manipulations, one noted scientist criticized molecular biology when he said, "Perhaps we're only 'genetics mechanics' today."[31]

Another broadly critical perspective was expressed when Paul Doty, long-time Harvard physical biochemist, was quoted as saying: "Because experts are burdened with too much knowledge, they have done poorly at predicting the future in science."[32] The reporter offered an interpretation of this view as "a warning to those who have become too enthralled with the powerful tools now available to molecular biologists. Thus, even the best and the brightest of these scientists may be running the risk of stagnation by burdening themselves with too much of but one kind of data."[33]

The Price of Ideology: Limited Approaches to Cancer Research

What difference does it make to our understanding of "life" to be guided by the basic question: How do the genes control everything in the cell? And, remembering Watson's descriptive and prescriptive injunction that "[h]ardly any contemporary experiment on gene structure or function is done today without recourse to ever more powerful methods for cloning and sequencing genes,"[34] what difference does it make to have research programs based primarily on recombinant DNA technologies that involve sequencing pieces of DNA and proteins?

A striking example of the consequences of this definition of molecular biology is found in the Darnell, Lodish, and Baltimore chapter on cancer—with important changes in the second edition (1990), to which I will refer. In the first edition (1986), the chapter treats cancer as a subject of molecular genetics. However, the authors point to poorly understood areas of cancer research. Comprehension of cell-to-cell interactions, cell surface biology, the ability of tumor cells to escape the surveillance of the immune system, cellular factors that affect cell division in the whole organism, and other growth controls was "stymied by complexity." Their explicit preoccupation with what the tools of molecular genetics have revealed about cancer is justified in the following way: "Enormous progress has been made in the areas that allow for genetic analysis because of the extraordinary power that molecular genetics has developed in recent years."[35]

This is another case of a misleading conflation of a successful methodology with an explanatory stance. The methodology of molecular genetics has, indeed, provided much information about a genetic analysis of cancer. But it does not necessarily follow that a genetic analysis is "enormous progress" or that genetic research is the best approach for understanding what causes cancer or how to prevent it.

Only in the chapter summary do the authors note some limitations: "This composite picture [of how a cancer develops] is certainly simplistic—for one thing, *it does not include nongenetic influences*—but it represents *a framework for future research,* and, we hope, for the development of new methods of prevention and therapy for this dread disease" (emphasis added).[36] The considerable omission of "nongenetic influences" is quickly countered by repeating the assertion that the genetic framework offers the (implied best) hope against cancer.

Furthermore, although the authors also point to the "enormous progress . . . made toward understanding how some external agents initiate cancer," they are more concerned about "how the initiating events affect the mechanisms that regulate cell growth and why abnormal cells fail to obey the rules of normal tissue organization."[37] Messages are mixed. Some information is given about cancer as it exists as a disease, but the emphasis is clear in the rest of the chapter, which is predominantly occupied by the genetic basis of cancer as it is studied in cell culture.

The authors sound almost apologetic about including a chapter on cancer in a book on molecular biology: "In focusing on a medical problem, cancer, this chapter may seem to have departed from the subject matter of the rest of the book. Cancer represents such a fundamental problem in cellular behavior, however, that many aspects of molecular biology are relevant to understanding the cancer cell."[38] That the authors justify a focus on cancer by casting it as "a fundamental problem in cellular behavior" suggests how tightly constraining are the boundaries separating molecular genetics and cell biology from the whole organism and from societal concerns at this time.

The provocative title of the last section of the chapter exacerbates the focus on genetics and cancer by asking: "Is Susceptibility to Cancer Inherited?" The answer provided: very rarely in humans. The authors' question as posed no doubt reflects widespread interest in the heritability of cancer. Yet the way the question was posed and highlighted tends to counteract the actual answer given. My question is: Why aren't other equally widespread concerns included, such as occupational hazards and environmental carcinogens? Even diet and nutrition are only mentioned in passing.

Consider the difference between thinking that cancer is primarily a problem of an individual's unfortunate genetic makeup, rather than a problem influenced significantly by cancer-promoting chemicals and irradiation in our air, water, and food, produced in good measure or affected by industrial processes and societal conditions and habits.[39] By viewing the problem of cancer through the abstract (but ideologically powerful) concern of heritability, readers are kept ignorant of the importance of factors we know contribute to cancer in various populations—and that we could change.

Several significant changes in the 1990 edition illustrate the problem of focusing on genetics in considering cancer, as well as how improvements can be made. The newer edition substitutes a section under "Human Cancer" entitled: "Rare Susceptibilities to Cancer Point to Antioncogenes."[40] The first sentence, while convoluted, stresses the *minor* role that genetic inheritance is thought to play in human cancer: "A corollary to the belief that multiple interacting events in our environment are the major risk factors for cancer is the belief that genetic inheritance plays only a small role in carcinogenesis."[41] The evidence immediately given is that "people who migrate to a new environment take on the profile of cancers in their new environment within a generation. For instance, when Japanese citizens move to California, they rapidly lose the oriental propensity toward stomach cancer, and soon show the occidental propensity toward breast cancer."[42]

The authors do not pursue this critically important example, however, since the focal point is genetics and oncogenes. Those of us concerned about the great increase in breast cancer in women in the United States over the past thirty years would have appreciated more attention to it—or encouragement for investigating this problem. This is but one of several examples in which an alternative to the focus on the gene is suggested but then dropped, neither developed nor offered as something the reader should consider pursuing.[43]

Then too, the discourse subtly discourages pursuing possible risk factors in human cancer in other ways, by focusing on "clear-cut," "hard evidence" as the only kind that deserves attention. After acknowledging that cancer research aims at changing the course of human cancer, the authors stress the "slow and frustrating activity" of identifying "clear-cut" risk factors, especially because scientists believe that "natural cancers result from the interaction of multiple events over time."[44] With that criterion, the only "successful endeavor" is "the identification of cigarette smoking as a crucial risk factor in lung cancer. A risk factor of this potency gives a clear indication of how to act to avoid lung cancer: avoid cigarettes."[45]

The language the authors use strongly suggests that we should not waste our time being concerned about risk factors that have not yet been as conclusively proven as cigarette smoking: "Animal fat is thought to be a risk factor for colon and breast cancer, and many viruses and chemicals have been correlated with minor cancers; however, hard evidence that would help us avoid breast cancer, colon cancer, prostate cancer, leukemias, and others is generally lacking."[46]

Readers are not exhorted to take up great challenges, such as getting the "hard evidence." Nor is it clear just what "hard evidence" is, as compared to other forms of evidence. No suggestions are given about avenues to follow to address this central problem in cancer research. *Nothing* follows to suggest that, *because cancer is multicausal and multistep,* we need to actively change the monocausal thinking to which we are trained (and to which we continue to train our students) in order to be more creative—and, ultimately, more successful—in our research. The authors provide little incentive to study complex interactions among different factors or economic priorities that influence our health, nothing more than a glimpse of epidemiological evidence that places the study of cancer into the context of society. Perhaps the best example of the consequences of this perspective is the refusal, until recently, of the National Institutes of Health to fund research on the effects of diet on breast cancer.[47]

Even with the second edition's changes, of the forty-two pages in the chapter on cancer, only the last two and a half focus on the "multicausal, multistep nature of carcinogenesis" and explicitly couple nongenetic influences with oncogene activity.[48] The changes made in the 1990 edition are significant (including more space on cell surfaces and other nongenetic components of the cell) and suggest increased awareness of the issues I raise, but they are not sufficient to construct a balanced view of cancer. Furthermore, the summary still ends by holding up the oncogene model as the hope for prevention and therapy against cancer.[49]

To say that "DNA alteration is at the heart of cancer induction"[50] is misleading both about the disease and what we can do to prevent or cure it and confuses mechanism with cause. What actually causes one person to get cancer while another does not is not the *mechanism* of oncogenes but the *effects on* the oncogenes (or proto-oncogenes). A straightforward statement to that effect could suggest that, if we want to prevent cancer, we have to change many things in our world, not just produce vaccines.

Here, an opportunity is lost to *integrate* the genetic and nongenetic influences conceptually, and to prepare minds to generate "a framework for future research" that encompasses, rather than excludes, the complex interaction of such factors. Instead, the text reinforces an artificial and polarizing boundary between genetic and nongenetic, which current science actually shows to be specious. Not only does our scientific understanding of this disease suffer as a consequence, but this form of the nature/nurture fallacy privileges the promise of a gene therapy approach to curing cancer, while ignoring the data that a large majority of human cancers are influenced or promoted by environmental carcinogens in our workplaces, in air, water, and food, in such cultural habits as sunbathing and tobacco use, and in our social conditions, such as poverty and stress. Rather than thinking of cancer as a disease not only of whole human beings but of society, the future researcher is encouraged by example to ignore major economic, political, and social forces that contribute significantly to this disease.

What is lost in minimizing "nongenetic factors" is critical. Many diseases are as multicausal and multistep as cancer, and focusing on eliminating only one risk factor or causal agent is narrow and counterproductive.[51] The representation of molecular biology in textbooks and journal articles generally keeps science separate from society (see chapter 8). Unless we challenge the current values and assumptions of molecular biology, our efforts to understand the workings of nature and the problems of disease will be partial, at best.

What must be changed is not only the paradigmatic polarization that occurs at all levels, constructing dichotomies such as genetic/nongenetic and nature/

nurture, but also the sociopolitical values and assumptions that structure and define molecular biology. Having examined the effects of cultural beliefs in molecular biology, we can imagine correcting some of those biases by eliminating gender-associated language and by bringing a historical understanding of the social construction of purportedly biological categories of gender and race to scientific studies of sex and sexuality, as well as race and class. We can present the parts of the cell as equally important in "life" and in "regulation" and avoid fallacies inherent in a reductionist worldview by understanding qualitative differences among the levels of organization we consider "life." And we can teach the history and sociopolitical significance of genetics, molecular biology, molecular genetics, and related subfields as an integral part of the construction of knowledge. Reconstructing systems of knowledge is not a simple task, but it is possible to imagine a fairly easy application of the dictum that feminist awareness functions as a necessary experimental control to eliminate pervasive and often unconscious gender bias. Other issues, such as that raised about the use of equality and inequality in genetics or the definition of "life" based on the gene, require more radical reconceptualizing of whole areas of biology.

Alternatives: A Beginning

An alternative understanding of genetics and the gene would balance many components and processes. DNAs, material complexed with it to make it function, RNAs and proteins produced, enzymes that modify the RNAs and proteins to a functional complex, interactions of that protein complex with other components as part of their functioning within the cell—all these and more constitute the entire process of life at the molecular level as we know it, with conceptual space for what we do not know.

Historical and philosophical studies of different frameworks within the sciences remind us of some alternatives found within our own scientific heritage. As Barbara McClintock and Lynn Margulis's research and lives illustrate, we can find alternative language, concepts, paradigms, and unifying principles in current science. But unless these counterexamples are brought forward to offset the predominant ideologies, they cannot serve as models for change.

Among the changes required to correct the imbalances I have delineated are explicit statements in textbook introductions, repeated throughout the books, that the molecular genetics approach is but one of several particular approaches that conceptualize "life." Scientists could point out that molecular genetics brings to the foreground the characteristics of growth and reproduction (the crystalline tradition), while other approaches (the fluid tradition) foreground maintenance of structures and functions amidst the flux of metabolism, energy, and physicochemical components—or the dynamic interactions of living beings with their environmental context.[52] Constant reminders should create more than a list of different views of life. Textbooks could interweave the differing perspectives at key points. Alternative epistemologies of "life" from other cultures could be compared to our culturally bound, Western scientific notions of "life."

Ruth Hubbard reminds us of Niels Bohr's use of "complementarity" in reconciling scientific characteristics of light as both particles and waves: "Classical physicists argued over which they really were. Bohr and the other quantum theorists asserted that they were both, and by complementarity Bohr meant that they were both at all times, not sometimes one, sometimes the other."[53] Countering a tendency to choose either one version or another, the notion of scientific complementarity will be particularly useful as a corrective to competing definitions of life. "Complementarity provides a fruitful model for integrating the different levels of organization we can use to describe living organisms. The phenomena we observe at the subatomic, atomic, molecular, cellular, organismic, or societal levels are all taking place simultaneously and constitute a single reality."[54] Both Bohr's notion of complementarity and feminist philosophy reject either/or polarities and hierarchies of differences.

Those of us concerned with refocusing molecular biology can work on such a transformation in many ways. These may include creatively exploring radical alternatives to language and concepts, developing new and more neutral language for balanced concepts, and looking for alternative language in existing scientific literature. I cite one example from a recent Nobel Prize address: Rita Levi-Montalcini received the Nobel Prize in Physiology or Medicine in 1986 (with Stanley Cohen) for her research, from the 1940s to the present, on a protein called nerve growth factor (NGF) that has many effects in a wide range of tissues. Her address describes the current understanding of nerve growth factor relative to its scientific history. In this case, the properties of the material itself—"a protein molecule from such diverse and unrelated sources as mouse sarcomas, snake venom, and mouse salivary glands [that] elicited such a potent and disrupting action on normal neurogenetic processes"—may have promoted a less reductionist and hereditarian view of the subject. In addition, it "did not fit into any conceptual preexisting schemes." Dr. Levi-Montalcini hints that she may also have resisted forcing NGF into the dominant framework derived from the study of bacterial viruses because her approach is more consonant with some traditions of animal physiology. She also suggests a view from the margin: "In spite of, or perhaps because of, its unusual and almost extravagant deeds in living organisms and in vitro systems, NGF did not at first find enthusiastic reception by the scientific community, as also indicated by the reluctance of other investigators to engage in this line of research."[55]

Levi-Montalcini's language suggests an alternative molecular biology, one that uses the latest techniques without having the results overpower and crowd out other components of interest. Her comments on "The Vital Role of

NGF in the Life of Its Target Cells," quoted above and below, suggest assessments relevant to the issue of foreground and background in science, paths not taken because of a dominant view defining what is significant to the near exclusion of other topics, methods, and approaches: "Between the two well-identified molecular entities of NGF and of its coding genes, which can be visualized as the summit and base of an iceberg, are several other possible intermediate forms of unknown nature and biological properties." Among the important questions their identification would answer is: "do alternative processing pathways result in the production of peptides endowed with different biological functions? Since the same peptides may undergo post-transcriptional or post-translational modification, the submerged areas of the NGF iceberg loom very large."[56]

Although she represents genes as the base of the iceberg, she grants equal status as "molecular entities" to both the protein NGF and its coding genes (an important distinction), along with unidentified intermediate forms. Levi-Montalcini highlights "the remarkable plasticity of the mechanisms controlling NGF gene expression."[57] The questions asked open outward onto a range of possibilities of unknown biological functions for the molecular entities. In more than one place in the text, she emphasizes the complexity of understanding these molecular entities in dynamic interaction with their specific contexts. Certain cell types "became the model of choice for studying the capacity of NGF to modulate phenotypic expression and the molecular mechanisms subserving this process. This, in turn, clearly pointed to the 'versatility' of NGF receptors and of their transduction machinery, whose message is evidently read and interpreted in different ways according to the cell type and previous cell history." That language emphasizes interaction and diversity in processing messenger RNA and explicitly acknowledges important effects of different local environments (different cell types), as well as the history of the particular cell. "The main causes of unpredictability of the findings reside in the intricacy of the new surroundings where NGF is moving—the CNS [central nervous system] and the immune system—rather than in NGF itself."[58]

Levi-Montalcini acknowledges certain *cell types* for their "contribution" to her scientific studies. "The 'priming model' which had the potential to give a molecular account of the very fast and very slow onset of neurite outgrowth . . . is an excellent example of the contribution of these latter cells to studies on the mode of action of NGF."[59] The values she expresses move those cell types away from a language of dead "systems" and toward a sense of living beings.

While a similar perspective that acknowledges complex interactionism is indeed included in textbooks such as *Molecular Cell Biology,* that more complex understanding is nonetheless organized and filtered through the dominant paradigm of DNA controlling life's activities. A more explicit and consistent description of molecular biology in nonhierarchical and interactionist terms is required to counter distortions deeply embedded in the language, paradigms, and unifying principles of current molecular biology.

In a search for alternative approaches to molecular biology and the development of alternative discourse, I found that the second edition of *Molecular Cell Biology* embodies some changes that illustrate an alternative direction from the gene's supremacy as the single most important component of the cell. One example is found in the treatment of developmental biology. In the first edition, the fourth and last part of the book is entitled "Normal and Abnormal Variations in Cells," and it treats four major topics: development of cell specificity, cancer, immunity, and evolution of cells. "Development of cell specificity" includes differentiation of cells and developmental biology. In the newer edition (1990), the authors explain one of the major revisions they have made in the book as a whole, "that, rather than isolating developmental biology and cellular differentiation into a separate chapter, we have integrated it with the other material in this edition. We did this because we recognized that this is not a text for teaching development but that gene control and cellular alteration must be seen in the context of developmental strategies."[60] Here an effort is made to place the discussion of genes and changes in cells in relation to developmental processes into a *context* of those processes. What difference does this change make? Does this particular change, which sounds on the surface like an improvement because it places "gene control" of development into a broader context, represent a real improvement in shifting the imbalance of genes to a more balanced representation of development?

The section on the development and differentiation of slime molds (see chapter 2) illustrates a change toward better balance in perspective. First, I will analyze what I see as problems (numbered [1]–[4] below) in the first edition's text with regard to the gene as the lens through which biology is to be understood. Then I will analyze the changes made in the second edition and identify what I consider to be improvements (lettered [A]–[C]). The following passage from the first edition addresses the question: "What causes a cell to differentiate along the spore or stalk pathway?"[61]

[1] This question of whether the cell environment determines cell fate or whether the features of the environment are produced by already determined cells arises again and again in developmental *systems*. [2] *Only the identification of the key genes that define the activities* of two different cell states and an analysis of *what controls the earliest expression of these genes* will shed light on such problems. No single problem of this type has been resolved yet.

No gene rearrangements such as deletion, inversion, or transposition have been identified as the basis for gene control during sporulation in *Bacillus,* yeast, or *Dictyostelium.* [3] Rather, differential transcription, differential mRNA stabilization, and differential translation seem the likely *molecular mechanisms* of differentiation. [4] In all three systems, a sequential *induction of gene products appears to lead the organism* through the sequential stages of development and differentiation.[62]

[1] The authors define causation as either/or: either the environment of the cell "determines" what the cell becomes, or the cells are inherently different and create changes differentially in their environment. This is an example of a misleading dichotomy of nature versus nurture, genetic versus nongenetic, as discussed in chapter 6.

[2] The authors propose that only gene analysis and the original cause of change in gene expression will give the answers. The strength of that assertion in the text is matched only by the challenge that no answer has yet been found to this important question, thus underscoring how essential it is to get on with the genetic analyses for the answer. This is a particularly clear example of subtle, yet undeniable, channeling of future research in one narrow and unicausal direction.

[3] This sentence is a clear example of granting active and hierarchical control only to DNA and not to cell processes and components. The three processes that seem to be responsible for the changes associated with cellular differentiation are termed merely "molecular mechanisms of differentiation."

[4] The language suggests that the "organism" is a passive vehicle for gene action. This not uncommon stance derives from talking about processes *of* the organism, *integral to* the organism (part of what the organism *is*), as if they were prior to and superior to the organism, for example, DNA directing the development of the organism. This separation of the parts from the whole, a useful method for investigating the physical living world, is conflated with how-things-are, an ontology, and assumes that one part is the horse that pulls the cart, so to speak. But development of an organism is an ontological whole; it can be said that the differentiation process of the slime mold or the embryo pulls the "molecular mechanisms" or the sequence of gene expression along. Either way, conceptually separating any part from the whole organism, even if we can physically separate them, and giving that separated part "control" misunderstands the dynamics of sequence, control, and change—and leads to conceptual distortions. We need to remember Bohr's notion of "complementarity," that activity is simultaneous at different levels of organization in living things.

In the second edition, the authors do not treat the issue of how spore and stalk cells differentiate. They do, however, describe the process of aggregation and the changes that occur in the single-celled amoebas as they aggregate. Aggregation involves coordinated movement of individual cells. The authors focus on the process of cell signaling that occurs during chemotaxis (attraction to a chemical), as the individual cells synthesize and secrete a pheromone, in this case, cyclic AMP (also known as 3',5'-cAMP).

[A] Proper functioning of this response requires a number of macromolecules: adenylate cyclase to synthesize 3',5'-cAMP from ATP; a cell-surface cAMP receptor; and secreted and cell-surface cAMP phosphodiesterases to degrade cAMP to 5'-AMP. The phosphodiesterases keep the extracellular hormone from building up to a level that swamps out any gradients.

Because cell-to-cell signalling by extracellular cAMP can extend over distances of only 10–100 μm, cells are attracted primarily to the cAMP released by adjacent cells. [B] Aggregation is initiated when random cells release pulsating waves of cAMP signals, which radiate outward from these "initiator" cells every 3–5 min; concomitantly, a pulsatile movement of the cells occurs inward toward the cell centers. The formerly homogeneous array of amebas rapidly breaks up into aggregation centers, each containing about 100,000 cells. . . . Specifically, the cAMP-occupied cell-surface receptor is a substrate for an intracellular kinase . . . , and [a] phosphorylated receptor cannot transduce the hormone signal and activate adenyl cyclase or cell movement. Over time, the phosphates on the receptor are hydrolyzed off, the receptor reacquires the ability to transduce the hormone signal, and the refractory period ends. [C] Further study of this receptor-transduction system will clarify the details of slime-mold differentiation and, possibly, the very similar signal-transduction system and its regulation in animal cells.[63]

In this selection, [A] the authors credit the requirement of many macromolecules in this complex process, as is often the case in actual molecular biological processes; [B] the text does not identify "initiator" cells as genetically determined or differently marked;[64] and [C] the future direction the authors suggest for research is not explicitly centered around genes, but seems to be cast in the complex framework of interactions among many different and important agents.

These examples only scratch the surface. The process of reconceptualizing the unifying principles of molecular biology and correcting imbalances and distortions in the dominant representations remains a major project for all those interested in how scientists represent the natural world.

Shifting the Balance: Creative Consciousness-Raising as an Avenue of Transformation

Many creative avenues can be explored across a spectrum that includes both conventional scientific frameworks and playful re-visioning. Collaborations among the overlapping groups of scientists, science students, and feminists seem promising. Playful and ironic disruptions of predominant paradigms and language can be liberating and fruitful. Ruth Herschberger's female-affirming descriptions of egg and sperm and female and male primates[65] and Susan Griffin's poetic explorations of genderized dichotomies[66] turn traditional science upside-down. When misunderstood as simply "the girls" doing what "the boys" have always done, such efforts may be viewed negatively.[67] I suggest that, for certain purposes, depictions of exaggerated gender polarities or gender reversals function as consciousness-raisers or heuristic devices to reveal cultural beliefs embedded in our assumptions about science and nature. Further, they can be freeing for an individual who has become aware of the constraints that bind our daily lives, our perceptions, and our imaginations.[68]

I am not suggesting gynocentrism (female-centeredness) as an accurate substitute for gender biases in this science, but I wish to encourage all of us to explore new metaphors for processes and relations among components in cell and molecular biology, including playful gender reversals. DNA might be conceptualized as a model of the spiral dance of life, a goddess metaphor.[69] Mitochondrial DNA could be highlighted for its role as an evolutionary fingerprint of the original mother of the human species.[70] Alternatively, RNA could be raised to high status by virtue of its apparent role as the original genetic material (the first mother molecule?), a perspective offered in the Darnell, Lodish, and Baltimore book:

> As we discussed briefly in Chapter 8 and will discuss later in more detail, RNA catalysis including self-cutting, self-splicing, and self-elongation and ligation virtually force the adoption of a central if not primordial role for RNA in precellular evolution. Not only can RNA perform catalytically on its own and as part of the snRNPs [small nuclear ribonucleoprotein particles] that process RNA but an ever-increasing number of additional roles for RNA in present-day biochemistry are being demonstrated, e.g., RNA serves the primer function in DNA synthesis; an RNA molecule is a necessary part of telomerase for adding the terminal DNA to eukaryotic chromosomes; 7S RNA is part of the protein secretion apparatus; a small RNA is part of the assembly structure for bacteriophage $\phi 29$; and, finally, dozens of small RNA molecules are known whose functions are as yet unidentified. . . . The case for the importance of RNA in precellular chemistry becomes virtually irresistible. DNA is chemically not nearly so flexible and even the deoxynucleotides are derivatives of ribonucleotides. . . . So virtually all commentators place RNA in existence before DNA during evolution of cells.[71]

Imaginative gynocentrism can be satisfying to feminists trying to make a more welcome place for women in the male-dominated sciences, and such playful superimposition of female symbols at the molecular level may be a useful strategy for breaking free of the power of dominant images and language in the sciences. While a feminist critical perspective reveals the potential harm of any form of gender ideology as a *model* for improving the status of women and the relationship of society and science, this critique does *not* mean that playing with gender-associated reversals or alternative gender arrangements is a useless or harmful endeavor. Traditions of satire, poetic license, and irony suggest otherwise, as the works of Herschberger, Wittig, LeGuin, Butler, and Haraway demonstrate.[72] In the face of the power of the prevailing paradigms and unifying principles of contemporary molecular biology, we must seek many sources of inspiration in order to disrupt and transcend the constraints of cultural bias in society and in science. This chapter has illustrated the price we pay if we do not risk these transformative efforts.

PART III

Science and Society in Molecular Biology

The important point is that science and technology have become such integral parts of society that scientists can no longer abstract themselves from societal concerns.

—National Academy of Sciences Committee on the Conduct of Science, *On Being a Scientist,* 1989

While Part II addressed gender ideology in the content of molecular biology, this third portion of the book looks at issues considered distinct from subject matter. Do current representations of molecular biology coincide with the view of the National Academy of Sciences Committee on the Conduct of Science in its 1989 publication, *On Being a Scientist?* Or do formal educational and scientific communications perpetuate consciously or not the belief that science and scientists are separate from societal concerns? And how do these representations of the relationship of molecular biology to society relate to feminist concerns for eliminating gender and related biases from the study of molecular biology in particular and science in general?

8

Molecular Biology

DISINTERESTED SCIENCE?

Politics and politicking preoccupied the first years of
the recombinant DNA story, but that phase, fortunately, is fast becoming history.
　　　　　—James D. Watson and John Tooze, 1981[1]

Is there such a thing as pure science? Well, there are
some things that are awfully close to pure and I don't
know how they can have any social relevance. . . .
　　　　　—Sheldon Glashow, 1989[2]

How is the social context of molecular biology presented? How are present and future scientists socialized to think (or not to think) about the beneficial and detrimental effects of their scientific field on people, other living beings, the environment, the biosphere? And what are the norms concerning scientists' responsibility to society?

The claim that science is a product of society and thus must be a reflection of particular concerns and interests (whether economic or intellectual) remains hotly contested. Scientists often react defensively to arguments that science is as much a creation of the values and beliefs of scientists as it is a reflection of what is "out there" in nature.[3] A belief in the inviolability of scientific objectivity may lead some scientists to consider critical science studies, including feminist critiques, to be thoroughly antiscience. But many others appreciate the need for students—whether as citizens or as scientists—to understand the interactions of science and society.[4]

Philosopher of science Helen Longino maps three positions held on debates about science and social values. One position believes that when "contextual values" (personal, social, and cultural) affect science, the result is "bad science." The second position, a strict social constructionist view (also called the strong program in the social study of science), has it that all of science is ideological and interest-laden. The third position reflects a compromise between the first two: that science is indeed laden with values and that as a consequence *some* scientific representations are incorrect, according to the internal rules of scientific inquiry.[5]

Longino, among others, has argued "not only that scientific practices and content on the one hand and social needs and values on the other are in dynamic interaction but that the logical and cognitive structures of scientific inquiry *require* such interaction."[6] Doing science depends on large-scale social cooperation and investment. Determining what research will be funded is based on judgments—whether by peer review committees reflecting the priorities of certain groups of scientists or by current commercial potential—that have long-term effects on conceptual approaches to major areas of study. For example, the biotechnological production of curative substances, such as growth hormone and antibodies, leads to research with an emphasis on a search for cures, rather than causes or prevention, of diseases.

The first portion of this chapter delineates a few of the key issues concerning the impact of molecular biology in society, particularly that of recombinant DNA technology, and analyzes major textbooks in molecular biology for their treatment of its positive and negative effects on society. The middle part examines scientific journals for prescriptions about the relationships between science and society and the role scientists play in this dynamic. The chapter ends with alternative examples, from a range of fields, that provide more inclusive and balanced pictures of the place of science in society and, conversely, the role of values in science.

Textbook Representations

Molecular biology today involves many critically significant and hotly debated issues that extend beyond the traditionally circumscribed boundaries of "pure science" into the realms of socioeconomics and politics. Some of the more important are the following:

—the promise of genetic engineering for curing diseases, clearing oil spills, producing biological products such as insulin more cheaply or more purely, and extending the range of agricultural crops; the subsequent ethical concerns around genetic engineering and gene therapy;

—the safety of recombinant DNA research, including concerns about the effects of altered organisms on the environment; issues about local and federal controls over such research as a consequence of the concerns about biohazards and ethics;

—the development of the biotechnology industry, with changing relations of academe, government, and private corporations, raising concerns about conflicts of interest, open access to scientific information versus secrecy in relation to legal ownership of profitable information and even bioengineered organisms, federal funding of research but private profits and high prices for products of biotechnology;

—creation of "big science" projects such as the multibillion-dollar Human Genome Project to sequence all the DNA in humans and other organisms, connected to concerns about funding sources and impact on funding of basic research projects.[7]

In information aimed at students, what stances are represented on any of these and other issues that are clearly science-and-society concerns? The field definition for molecular biology in *Peterson's Guide* (see chapter 2 and Appendix B) includes a lengthy reference to the recombinant DNA revolution and its central place in the growth of the biotechnology industry. The description is uncritically positive, focusing pragmatically on opportunities for positions in both science and business. Striking by its absence is any reference to opportunities for applying one's training in molecular biology and genetics to the legal, ethical, or environmental aspects of biotechnology and recombinant DNA, suggestions that would be equally practical for future career considerations. Missing, too, are negative implications that require attention.

The authors of one of the major introductory textbooks, *Molecular Cell Biology*, end the preface to the newer edition by acknowledging the significance of biology to the lives of all citizens today: "We believe that a comprehension of modern biology is needed both by those who use biological concepts professionally and by the general public, who increasingly will be faced with decisions about integrating new biological understanding into the fabric of their lives." The authors then state their intention to "help both groups to better comprehend the revolution in understanding of living systems that is being generated by research laboratories around the world."[8]

The text, however, undercuts this laudable goal with a one-sided, positivist philosophy about the value of science to society: that scientific progress accomplished as rapidly as possible is in itself good for all in society. The epigraphs chosen to set off the textbook's opening to the introductory chapter on the history of molecular cell biology capture the book's unbridled enthusiasm for reductionist progress. First, François Jacob's *The Logic of Life* is quoted: "The aim of modern biology is to interpret the properties of the organism by the structure of its constituent molecules." Then, James Watson's classic molecular biology text asserts that "we have complete confidence that further research, of the intensity recently given to genetics, will eventually provide man with the ability to describe with completeness the essential features that constitute life."[9]

Indeed, universal and self-important "man" draws attention to the issue of his relationship to nature. The "complete confidence" and "completeness" of this scientist's attitude epitomizes a belief in "man's" total power to work nature for his benefit. Such claims are repeated in grandiose statements about the promise of science to fix society's major ills, as I illustrate later in this chapter.

The message about the usefulness of the molecular genetics approach to scientists and to the study of nature is quite clear in the book's celebration of "the determination of today's biologists to carry the spectacular successes of the 1950s and 1960s—the discoveries of the structure of DNA, the roles of RNA in protein synthesis, the genetic code, and the nature of gene regulation in bacteria—into studies of the cells and organs of higher organisms, including human beings." Untempered enthusiasm characterizes the tone of the book, as the authors describe how "this group of techniques . . . collectively called molecular genetics became the dominant approach to the study of many basic biological questions."[10]

"Spectacular successes . . . revolutionized . . . a new era . . . fantastic strides forward . . . great optimism . . . few problems will remain unsolved . . . only the identification of the key genes will shed light on such problems . . . power and success of the new technology . . . raised high hopes . . . will bring many benefits to mankind."[11] Irresistible—who could be against such progress and so many humanitarian benefits? What the textbook lacks is some hint that scientific "progress" has not automatically benefited large numbers of people, particularly those lacking resources and power. Missing is some reference to what society has learned about the destructive aspects of "astonishingly rapid" change. In the larger sense, the book lacks a societal context in which to consider a range of perspectives other than that science and its consequences are intrinsically good. If we ask "For whom is this assumed good true?" we break down the illusion created by universal, purportedly objective boosterism and bring into focus the different faces of the public, including the scientists themselves. What is certainly lacking is evidence of self-reflection, of an understanding of the destructive consequences of that positivist assumption itself.[12]

The total absence of any reference to the controversy about potential biohazards and ethical problems raised by recombinant DNA technologies is a glaring example of the suppression of societal context in science. At least one of the authors was among the scientists, working with the then new DNA technologies, who in 1973 initiated discussions and a temporary moratorium on such work. With the ability to cross species boundaries by inserting genes (such as those from SV40, a virus that can induce cancer in certain animals) into the bacterium *E. coli*, which resides in the human gut, scientists recognized potential hazards from such research and organized meetings to agree on safety guidelines to minimize potential biohazards. Quickly, however, the lay public entered the debate. At the local level, in Cambridge, Massachusetts, the flamboyant mayor and the city council held public hearings to decide whether to grant building permits to Harvard University and MIT for high-level containment facilities for potentially hazardous research. More disturbing for the scientists were Senator Ted Kennedy's efforts at the federal level to pass legislation regulating this kind of research. Ultimately, the National Institutes of Health passed guidelines that could be enforced only after the fact by canceling federal grants to individuals and academic institutions shown to disregard them; industry, increasingly involved in this kind of biotechnology research, is not bound by those guidelines.[13]

By 1978, scientists had successfully lobbied against the federal bill, and fears about potential biohazards were, in the prevailing view, laid to rest by pointing to the absence of illnesses associated with the research. The debates among scientists became acrimonious; professional and personal relationships were permanently severed.[14] Furthermore, many research scientists moved into industrial research and development, often through the creation of biotech companies as subsidiaries of major pharmaceutical corporations. This change dramatically challenged the claim of "disinterest," as economic interests and ties to industry became more apparent, marking a significant shift in the relationship of this area of biology to the institutions and structures of scientific production.

Clearly, the lens of recombinant DNA technologies was polished in controversy—about risk/benefit assessments (by whom and for whom), about scientists' control over their science, about the role of public interest in directing scientific research, about the relationship of academic scientists to industrial interests, and about ideology shaping science.[15] Although textbook authors have a right to their view of that controversy, erasure of its existence is intellectually dishonest. The new generation of molecular biologists is shielded from the fifteen years' worth of arguments about the safety and ethics of recombinant DNA research and its relationship to genetic engineering. And they are taught that such considerations are not science.

Watson's newest edition of *Molecular Biology of the Gene* provides an instructive, but inadequate, alternative to the treatment of science-and-society is-sues. Watson et al.'s preface exudes a similarly unwavering certainty about the importance and benefits to society from recombinant DNA technology:

> It is only in this fourth edition that we see the *extraordinary fruits* of the recombi-nant DNA revolution. Hardly any contemporary experiment on gene structure or function is done today without recourse to *ever more powerful methods* for cloning and sequencing genes. As a result, we are *barraged daily by arresting new facts of such importance* that we seldom can relax long enough to take comfort in the *accomplishments* of the immediate past. The science described in this edition is *by any measure an extraordinary example of human achievement.* . . . DNA can no longer be portrayed with *the grandeur it deserves* in a handy volume that would be pleasant to carry across campus.[16] (Emphasis added.)

Watson *does* make reference to the recombinant DNA controversy in the preface, but how he treats the issue is noteworthy. Readers are first given a long buildup to the remarkable achievements presented in the current text-book. In that context, a cautionary note is raised and then dropped:

> As this new era of molecular biology began, however, there initially was widely voiced concern that recombinant DNA procedures might generate dangerous and pathogenic new organisms. It was not until after much deliberation that in 1977 the cloning of the genes of higher organisms began in earnest.[17]

Readers are simply reassured, without any specifics, that serious concerns about safety were addressed expeditiously and responsibly by scientists so they could get on with the important business at hand, the "extraordinary example of human achievement."

The question remains, however: Should those concerns about safety be dis-missed so easily? This has been and continues to be a complex point of con-tention amongst scientists and citizens. I am not advocating one view as the correct one; rather, I am highlighting the ignorance fostered by these text-books about the significance of these issues—and the lack of preparation of-fered to students for grappling with these concerns as citizens and scientists. In a collection of widely divergent essays published by the American Associa-tion for the Advancement of Science Series on Issues in Science and Technol-ogy, Sheldon Krimsky identifies as the central controversy in the 1970s such human safety concerns as new disease-causing bacteria. But, he adds, other neglected aspects of manipulating genes are equally significant, including "an expanded view of biohazards to include ecological disruptions from modified soil bacteria, for example; the ethics of human genetic intervention; the ques-tion of justice in the distribution of the fruits of biotechnology; the setting of priorities for the applications of the technology—for profit or human needs; the possibility of using the new techniques to create biological weapons."[18] This range of issues clearly deserves consideration—as part of what is taught as molecular biology.

Both of the major textbooks stress only the positive promises of recombinant DNA technology, without reference to potential hazards and the difficult problems of how scientists and the public are to decide what is hazardous and how to proceed with known and unknown risks. Further, they do it without raising students' awareness of the ethical, political, and economic issues generated by the application of recombinant DNA technologies in industry and in genetic engineering. The inescapable conclusion from examining these two major textbooks in molecular biology is that the authors, intellectual leaders in the field, have fallen short of their expressed goal of helping scientists and citizens alike to make informed "decisions about integrating new biological understanding into the fabric of their lives."[19]

The acrimonious debates over recombinant DNA research—with the sudden threat of federal and local control over basic research in molecular biology—have left their mark on all participants. Those who agree with the decisions of the late 1970s to leave the governance of science to the scientists (and industry) and to loosen the guidelines for recombinant DNA research, assuming that any biohazards are under control, are not likely to dredge up the debate for public display once again. Nevertheless, everyone who uses these texts should know that such issues have received attention but are not totally resolved, and readers should be directed to the range of perspectives, arguments, and subsequent analyses of the issues. How, indeed, are scientists to be prepared for the very issues they are forced to face in a climate of increased concern about accountability of professionals in society? And how are members of the public to become knowledgeable about the intersections of scientific study with their lives, if not by learning about the history of such controversies and, perhaps, learning from them as well?

I experienced an analogous lost opportunity in my own prestigious education in science in the 1960s and 1970s. Nowhere did I learn about scientists' involvement in the development of the atomic bomb and their futile efforts to prevent the United States government from dropping the bombs on Hiroshima and Nagasaki. I believe that scientists of my generation and future generations have much to learn from the history that shows the naïveté of those scientists, brilliant as they may have been, who earnestly sent their petitions and their representatives to President Truman, proposing a demonstration of the destructive power of the bomb without killing human beings. Information released in the 1970s shows that the decision to drop the bombs had been made by Truman well before then and that there was no consideration of changing that decision.[20] Courses in history and social studies of science that include this and similar topics are generally absent from the science curriculum because they are about "society," not "science." If education about the politics of science does not take place in the science classroom, we run the risk that the majority of science students will perceive politics as irrelevant to science, when exactly the opposite is true.

Coming from a long tradition of education that obscures the connections between society and the science it produces, most scientists are socialized to what they think is an *a*political stance in science. The apparent absence of "politics" from the textbooks and the curriculum can be understood as a product of a tradition that I suggest must and can be changed.

But how are such traditions maintained? Who benefits from such denial and masking of the actual political dimensions of science? What can we find out about the process of preserving purposeful or unintentional ignorance of the politics of science? Susan Wright has analyzed three conferences that produced what was called consensus on the safety of recombinant DNA research. A reading of the transcript from the conferences reveals that the scientists' evaluation of potential biohazards was shaped by political concerns about who should control scientific research—and by a set of assumptions that circumscribed consideration of certain important issues (for example, the question of possible entry of new gene combinations into organisms in the environment).[21]

Reading Wright's narrative of conference participants' discussions of ethical concerns is frightening:

> [T]he sense . . . that biomedical research was threatened came increasingly into focus. When several people noted that other aspects of biomedical research might pose hazards as serious as those of the new biology, they were warned that scientists must be careful not to stimulate the spread of regulation to other research fields. "Science," someone announced, "is under very serious attack." "But where is the attack coming from?" it was asked. "From ourselves," came the answer. "One has to be very careful about the tack one uses and should not say, 'Well, gee, we have been doing much more dangerous experiments for years.' That's murder! You have to use a very positive approach." In the same vein, someone else (or possibly the same person) warned that:
>
> "we have a serious political disease . . . you have to be careful in these arguments that you don't spread it to other people. The big danger about the argument 'But look! something else is much more dangerous than what we do already' is that the 'something else' all of a sudden gets in with a big bag of red tape at the very least."[22]

Wright continues:

> When someone at this point attempted to make scientific distinctions about hazards, they were told that the political dimension had to be emphasized. . . . The issue was not *whether* the "epidemic pathogen" argument was technically acceptable but *how* it should be used politically. As someone summarized the sense of the group at the end of the morning session:
> "I think [the problem of convincing the public] is what you have to deal with. It may not mean a thing, but it is very easy to do. It's molecular politics, not molecular biology, and I think we have to consider both, because a lot of science is at stake."[23]

Scientific deliberations shifted away from evidence and speculation about potential biohazards as a consequence of gene splicing toward what the public should be told in order to preserve scientific research as unencumbered as possible from regulations and red tape. Wright's documentation and analysis reveal that scientists' political concerns about who should be able to tell them what science they can do—and in what time frame—produced a restricted discussion of biohazards, which ultimately closed down debate about the safety issue before all aspects of it were addressed.

The picture that Watson and Tooze present in their oversized "documentary history" of the recombinant DNA controversy differs sharply. In *The DNA Story*, the authors create a picture of the scientific process as one in which scientists eschew politics in order to get beyond such time wasting to the *really important and beneficial* work of doing "science," pure and unfettered by political considerations.

Excerpts from the last section of the book, which provides the scientific background of the debate about recombinant DNA, reveal an effort to sound evenhanded about the way scientists raised the issue of hazards to determine whether recombinant DNA work should be delayed until all the proof of its safety was in:

> The only question was whether to move ahead as fast as possible or to try to invent methods that would reassure the worriers without straightjacketing most future explorations of recombinant DNA. As the documents we have reproduced in this book show, although the decision on which way to proceed could initially have gone either way, the cautious approach prevailed.

The tone shifts slightly in the next sentence:

> The exploitation of recombinant DNA research was subjected to regulations and massive bureaucracy that only now are being lifted.

The anger rises:

> At a cost of what must have been many millions of dollars, the recombinant DNA debate was pursued worldwide.

Then it bursts forth:

> Delays and frustrations were forced on those wanting to get on with what were bound to be exciting experiments.

The epilogue sums up the thesis of the authors, presented as assertions of fact revealed through the purportedly careful history and documentation of their book:

> For eight years recombinant DNA research was placed under constraints that were crippling. . . . That so much new and totally unexpected has been learned since 1973 about the structure, organization, and expression of genes in higher organisms—despite all the obstacles put in the way of recombinant DNA experiments (recall that in the fall of 1976 clones of yeast and fruitfly DNA were ordered to be destroyed)—is the best testimony to the power of these methods.[24]

Further, the primary justification for forging ahead is the extraordinary contribution of recombinant DNA research to *save* the field of molecular biology. This assumes that the field needed saving and that only recombinant DNA technology could do it by its extraordinary power.

> Recombinant DNA is no ordinary technical development; in combination with new ways of sequencing DNA it is a tool of enormous power. It came at a time when the conventional genetic and biochemical techniques . . . were becoming inadequate. . . . Without recombinant DNA methods, further progress . . . seemed destined to be painfully slow, disappointingly meager, and very expensive. Recombinant DNA rescued the field from a future of gradual atrophy.[25]

Note the descriptions of the supposedly declining field: "inadequate," "painfully slow," "disappointingly meager," "gradual atrophy,"—language reminiscent of male sexual inadequacy.

According to Watson and Tooze, not only did gene splicing rescue the field of molecular biology from decline and impotence, its other key contribution was that "it offered the prospect of new industries based on molecular biology."[26] Here is an unabashed statement by one of the most powerful leaders of molecular biology today about the purpose of science in our society and the uses of "powerful" scientific methods; according to the authors, a major purpose of science is corporate development. Left unsaid is that the biotechnology industry was built at the expense of taxpayers, who footed the bills for "basic research," while financial benefit has gone to a few scientist-entrepreneurs and the backers of public stock offerings on the new companies.[27]

The epilogue continues to hark back to the horror of "unnecessary restrictions" and delays. "Remember," the authors admonish us, "that in August 1976, *The New York Times Magazine* devoted several pages to a feature article calling for a worldwide moratorium on all recombinant DNA research." And who is to blame?

> The *disastrous way* in which the issue had captivated the public's imagination [not rational thinking based on evidence] owed more than a little to the *strident* way the *molecular biologists* first sounded an alarm about the possible risks of recombinant DNA.[28] (Emphasis added.)

What is the clear message to students of biology? To avoid "disastrous" consequences for science and for the good of society, do not become the "strident" scientist who sounds an alarm; do not involve the public (its "imagination" or even its "knowledge") in matters which scientists are most capable of addressing without public interference. And, in case the reader missed the point, the epilogue ends with this statement:

> Politics and politicking preoccupied the first years of the recombinant DNA story, but *that phase, fortunately, is fast becoming history.* This book is our *epitaph* to that extraordinary episode in the story of modern biology.[29] (Emphasis added.)

Scientists, in this view, are not to be political. They are to leave politics to the unsavory past. Everyone learns that the controversy over the safety of recombinant DNA research was a "disastrous" event, never to be repeated.

The ironies in this are many, including scientists' concern during the debates for "protecting 'free inquiry'," referring only to the fear of government red tape and regulations of the research. In actuality, "free inquiry" has been severely hampered by the "new industries" applauded by Watson and Tooze. Substituting patents and profits for the goals of scientific inquiry in academe necessarily narrows the exchange of public information in recombinant DNA research. The ultimate irony is that Watson and Tooze's book itself is a major political tract that obscures its politics, presenting an inaccurate and unbalanced "documentary" about the controversy, and promoting one view of science, an exaggerated positivist view: Scientific progress is good for society, good for everyone; while science is itself neutral in its values, it brings no negative consequences to speak of; scientists can and should regulate themselves and their work because they have the good of society in mind and also the specific knowledge required to make informed decisions about science. Watson has been the major promoter of the Human Genome Project, a multi-billion-dollar project to sequence the genetic material in humans (which humans?). This project is highly controversial in its extreme expense, its cost-effectiveness, and its broad utility. James D. Watson is one of the most overtly "political" scientists in the world in the sense of exerting power over a portion of the scientific arena. His obfuscation of politics and politicking in science is pure propaganda.

The messages in *The DNA Story* cannot be trivialized by saying that such a book would not be read by many people or by science students. To do so would be to ignore Watson's importance in the history of molecular biology. It is possible that more American adults know Watson's name than any other biologist's after Darwin and Mendel. He is the most famous Nobel laureate in molecular biology. Further, while John Tooze may not be a household word, he is well known to advanced students of molecular genetics and animal virology.

In contrast to this large book, written by the most well-known molecular biologist and a respected colleague and published by W. H. Freeman, Susan Wright's article is to be found in *Social Studies of Science* (a journal *not* found in all university libraries). While alternative perspectives such as Wright's are not suppressed in the academic literature,[30] they are not nearly so widely read by scientists or assigned to science students as Watson's work and views.

Adding these propagandistic messages to the lack of textbook information about scientific-political controversies about the technologies that are proposed as the core of the new biology, we can speculate about the consequences for various students. Those who don't know much about the controversies would pick up that "politics" is not the domain of scientists; and that the scientists who do raise those concerns are being irresponsible to science and to the public. Further, it means that any time such issues *are* raised—by peers or by teachers—students have been predisposed to see scientific political controversies as irrelevant, a judgment easily extended to the people raising them. And who might those people be? Feminist scientists (who are dispro-portionately women) may be more apt to address science-and-society issues openly within their classroom than nonfeminist or antifeminist scientists (who are disproportionately men). Scientists who are overtly politically active (disproportionately radical or leftist) would also be more apt to raise such issues. The messages from Watson and from major textbooks reinforce a con-viction that politics are absent from the real or pure science that scientists do for the good of society—and that efforts to analyze the invisible political con-tent in the organization, production, and content of science are antiscience.

Another key assumption embedded in this scientific literature is that no *mention* of politics means *no politics* are present. This assumption operates across disciplines in academe, evident in charges that feminist, Afrocentric, and ethnic scholars are injecting politics into the curriculum and forcing a "politically correct" position on others.[31] Such a charge ignores or denies the politics of a curriculum that represents and lauds only the exploits and inter-ests of a small portion of the population: privileged white males. Most of us have been trained in a formal education with one acceptable position, that of the dominant Western, capitalist worldview, which is also sexist and racist. However, the studied *invisibility* of "politics" in our traditional curriculum means that the "great men of the Western world" and their supporters can claim objectivity and universal criteria instead of recognizing or acknowledg-ing the particular and narrow interests served by those teachings.

The impact of such messages for students who *do* know something about the recombinant DNA controversies and other openly political issues in sci-ence may be that they learn not to raise such points in the science classroom and that their political concerns in relation to science must be pursued sepa-rately from their education as scientists. Thus, to be taken seriously in science, most students learn to separate their interest in science-society issues from their science. Students who raise the issues in the classroom or the lab or the assigned paper often run the risk of exacerbating the conflicts they experience about the boundaries of science and politics.[32] For feminist students, for ex-ample, it may be more practical (albeit unfortunate) in a given academic set-ting to keep their feminist concerns separate from the science classroom. Some women's studies students who have experienced support either in women's studies classes or with feminist advisors or within their feminist community have courageously tried to raise such issues in biology or other science classes, with mixed results. Since the politics embedded in the status quo are rendered invisible in much of academe, the inference is often drawn that raising issues about power inequities in science is *injecting* unnecessary politics into science.

If my insistence that patriarchal Western values underlie molecular genet-ics seems a concoction of a feminist imagination, we have only to return to James Watson and John Tooze. Two other statements from the prologue of *The DNA Story* deserve our attention in this analysis of representations of molecular biology and science in general. The first statement presents a clear masculinist and racist philosophy of science:

> But from the start [of discussions about potential hazards of recombinant DNA research] we knew that no one had *concrete facts* by which to gauge these scenarios of possible doom. So perhaps we *best* proceed in the fashion of the *past 500 years of Western civilization,* striking ahead and only pulling back if we find the *savages not of normal size but of the King Kong variety,* against which we have no chance.[33] (Emphasis added.)

According to Watson and Tooze, serious scientists do not waste their time with speculation. The tradition of Western imperialism will show scientists the way, "striking ahead" with the power of the past half millennium of unstoppable Western civilization. With issues of gender, race, and class now being addressed, it cannot escape our notice that the central metaphor chosen by Watson and Tooze involves other legacies of that period of Western civilization: racism and imperialism. First, the authors equate "savages" with the bad (uncontrollable, unexploitable) side of the scientific exploration of nature. Secondly, the equation of "savage" with a giant gorilla is grotesque, but by no means unique in Western culture. What does it mean for a field of science when one of its most powerful spokespersons expresses racism with equanimity? And when his community—including editors—allows it to go unchallenged? Rarely have the analogies among Western colonialism, racism, and scientific discovery been so clearly revealed.

The second statement is part of a brief summary of the rise and utility of genetic engineering. Ethical and safety concerns are dismissed: "Although previously there had been numerous *speculative scenarios* for altering our genetic compositions, not one had a ring of *plausibility,* and for all practical purposes they were *indistinguishable from science fiction,* a genre upon which neither one of us had *ever wasted any time.*" The power to control nature emerges triumphant: "But with the new recombinant DNA tricks the genetic engineering of microorganisms, and later of higher plants and animals, would help to shape the world of the future. *Without doubt,* molecular geneticists now had the power to alter life on a scale never before thought possible by *serious* scientists."[34] (Emphasis added.)

This view caps the edict that science is not and should not be political, a view that places blinders on the political content of all science. Taking an ahistorical stance as well, the authors do not even credit science fiction for presaging technological developments, such as rockets to outer space and computers, much less recognize the value of science fiction as an arena for exploring societal values interlinked with scientific progress.

Most significantly, the epistemology embedded in these statements is characterized by a mutually exclusive duality of scientific fact as truth, on the one hand, and, on the other, science fiction and fantasy, creating distinctions between what is "true" and what is merely imagined and invented. This artificial dichotomy denies the possibility of the social construction of science.[35]

Scientific Journals

What are prevailing attitudes in scientific journals that address some science-and-society issues? First I examine *Cell,* one of the foremost publications in molecular biology (but with little space given to science-society topics) and then *Science,* which devotes sections of its weekly publication to such concerns.

Cell. For scientists in the field of molecular biology, *Cell* is a must-read, not only for reports of original research, but also for its "mini-reviews," short reviews of hot areas of developing research. Typical of most scientific journals, *Cell* usually does not publish editorials or letters. Every few months, however, an occasional editorial by Benjamin Lewin is printed. In the 19 May 1989 issue, Lewin's editorial, "Travels on the Fraud Circuit," addresses charges of scientific fraud leveled at a scientist in David Baltimore's laboratory at the Whitehead Institute at MIT. Lewin redirects attention away from the fraud itself to the issue of whether science should be "further regulated" and, if so, how and by whom.[36] He expresses concern that, as attention to fraud in science has increased, its incidence is taken more and more for granted, so that meetings now address what scientists' response to it should be.

Only a very small part of Lewin's editorial pertains to the accusations that a paper published in *Cell* may contain unsupported data. Indeed, the editorial is immediately followed by a rare letter to the editor, written by three of the authors of the questioned paper. The authors provide additional information about specific methods employed in the original work, justifications of the absence of some data, an admission of "a few clerical errors" found in a key table, and the addition of a corrected table. Lewin strongly criticizes what he sees as a biased process stemming from the zeal of one Walter Stewart, a federal administrator-scientist, whose power to inflame a congressional committee (the Subcommittee on Oversight and Investigations) has resulted in its questioning the conclusion of an NIH committee that fraud was not involved. The message to readers of *Cell* is clear, as is the specter of government intervention into research:

> An irony is that future hearings could damage the ability of the scientific community to handle such matters; who will be willing to serve on committees of investigation if they themselves are the next to be investigated?[37]

Regarding the particulars of that case, Lewin asserts, "The affair of this paper should be closed with the publication of the latest statement from the authors."[38] However, the issue was not closed, and two years later another NIH panel concluded that the whistle-blower, then postdoctoral researcher Margot O'Toole, was justified in her charge that the data published in *Cell* did not match the raw data of research notebooks. (By that time O'Toole had lost her job and her house.) Consequently, David Baltimore is reported to have said that Thereza Imanishi-Kari, the accused researcher, should confess and re-tract the paper, although a number of other scientists subsequently protested her treatment.[39]

That the paper turned out to contain fraudulent data may not affect Lewin's or others' views that scientists and institutions should be allowed to regulate themselves. Lewin proposes that the National Academy of Sciences establish mechanisms for resolving disputes. "It would be wise to develop such a mech-anism within the scientific community before it is imposed from the outside." While Lewin concedes that institutional responses to allegations of fraud or misconduct have been "poor, in some cases, disastrous" and that scientists have allowed this vacuum to exist, his starting premise is that government interference is bad for science and scientists. His only example to support this contention is a telling one: the debates about genetic engineering. From that experience, he says, scientists learned that:

> it is difficult for those outside the scientific community to distinguish one scien-tist purporting to offer impartial advice from another. The best way to prevent those with their own agenda from exploiting the present issue would be the existence of a general (suprainstitutional) mechanism to resolve disputes. . . . It would be wise to develop such a mechanism within the scientific community before it is imposed from the outside.[40]

Lewin offers no suggestions about how that purported state of affairs—where one scientist's claims cannot be distinguished from another's—came about, nor does he state whether scientists and educators have a role to play in changing that. His editorial is a call to scientists to protect themselves and their colleagues from meddlesome interests that are assumed to be signifi-cantly different from those of scientists.[41]

In this rare instance when *Cell* directly addresses an issue of the relation-ship of science to society, the message is to keep science separate from the concerns of the public and the government. This view seems to contradict another rare publication of letters, concerning the issue of increased scientific secrecy in molecular biology, secrecy arising from research by private compa-nies. The letter writer's solution is to recast the problem and turn scientists' obligations upside-down with regard to sharing research materials and infor-mation. Rather than addressing the origin of the problem—secrecy in corpo-rate scientific research is at odds with the ideals of scientific freedom of communication—he proposes that scientists not employed in industry simply scan patent announcements to determine whether information on a topic of interest is available. In a stand apparently different from the previous exam-ple, the author assumes that science (usually considered pure research, rather than applied work) and society (here, industry) must accommodate each other.

A common philosophy unites the two views highlighted in *Cell:* that "coop-eration, not compulsion" guide the distribution of research materials, "respecting individual freedom in the pursuit of science." Any kind of intervention (even guidelines) in science seems to be anathema to the scientists whose views are represented in the major literature. "Individual freedom" becomes synonymous with "free enterprise," as the biotechnology industry's restrictions on open communication supersede the norms of scientific communication. Beliefs in free enterprise and American individualism also ignore the ways that government, industry, and foundation funding direct and shape scientific research. Most ironically, this view refuses to acknowledge the power exercised by a few individuals, such as the editor of *Cell,* who, like the editors of *Science* and *Nature,* exerts dramatic control over what is published in his influential journal, which raises serious questions about the use of peer review and balanced discussion of certain controversial scientific issues (see below).[42]

Science. Daniel E. Koshland, Jr., the editor of *Science,* has made his philosophy of science clear in his weekly editorials:

> [T]he goal [of *Science*] should be to present as close to the unvarnished truth as it is possible to achieve. As the interface between science and society becomes more important and complex, there must be a journal that is widely respected for presenting, with perception and candor, the developing and changing data and theories of the controversial issues of the day. Evangelism for the truth must be based on the humility that there is not omniscience at the frontier.[43]

"Unvarnished truth . . . interface becomes more important and complex . . . the developing and changing data and theories [not changing politics] . . . [still] at the frontier": the tug of overt politics, the encroachment of society on the doorstep of science, these are palpable in Koshland's pronouncements. His response to the perceived relationship of science to society and recent changes has been to advocate that scientists try to control the interface, retaining traditional beliefs in "unvarnished truth" and the apolitical nature of science as it exists today and has existed since the inception of Western modern science.

Articles in *Science* frequently hide their politics. "Sources of Human Psychological Differences: The Minnesota Study of Twins Reared Apart," for example, a report on the performance on psychological and IQ tests of monozygotic (identical) twins raised together and raised apart, claims to corroborate smaller studies and to be able to assign a percentage (70%) of the variance of IQ to genetic variation. The abstract to the article sounds somewhat balanced, moving the influence of genetics on psychology to a less direct (though unspecified) mechanism:

> It is a plausible hypothesis that genetic differences affect psychological differences largely indirectly, by influencing the effective environment of the developing child. This evidence for the strong heritability of most psychological traits [such as religiosity and traditionalism], sensibly construed, does not detract from

the value or importance of parenting, education, and other propaedeutic [instructional] interventions.[44]

The "many family, twin, and adoption studies" referenced in the paper have been shown to suffer from serious conceptual and methodological flaws. For example, many twins raised apart are actually raised by relatives with similar socioeconomic class, giving the lie to the assumption that the environments of the genetically identical twins are radically different. Furthermore, the original twin studies that created the field of genetics and IQ have been shown to be purposely fraudulent. Eugenics supporter Cyril Burt (later knighted) was so convinced that heredity determined IQ and that lower-class people were inherently less intelligent than the economically privileged that he invented much of his data, going so far as to make up the research assistants who he claimed had gathered the data. This blatant example of conscious fraud on the part of a pillar of the scientific community is no less shocking than the realization that the evidence of fraud was clear fifteen years prior to the discovery, but no one recognized it. Further, despite consistent methodological and conceptual inadequacies (such as inadequate size of sample, unrepresentative sample, skewed subjective judgments, selective adoption, and common environment of "separated twins"), the same kinds of studies with the same kinds of claims are still published in scientific journals like *Science*.[45]

Behind author Bouchard's talk of "intervention" to change IQ performance lies the belief that the "genes sing a prehistoric song that today should sometimes be resisted but which it would be foolish to ignore."[46] This is the biological determinist dogma of human sociobiology, shown by critics to be based on incorrect assumptions, questionable data, and inadequate interpretations of evolution. However, none of these critiques are suggested in either Bouchard's article or the journal's commentary. In fact, Bouchard's work, supported by the ultra-rightist Heritage Foundation, is included in the first summary featured in "This Week in Science." In a highlighting of plans to map the human genome (the theme of that *Science* issue), the summary asks: "Just how important is a human genome for individual development?" and answers with Bouchard's work: "Bouchard et al. summarize the results of a continuing study of identical human twins who have been reared apart; this study shows the strong role played by the genome in IQ development." Then, Koshland's editorial is cited.[47]

Indeed, Koshland states in his editorial, "Today research in the neurosciences is flourishing" and then points to eight articles in the issue, from "molecular manipulation of ion channels to a study of primate behavior to a study of human twins." Although the human twin study has nothing to do with the neurosciences, casting it that way legitimizes it, as if it were an empirical and precise "science" such as the ion channel studies in molecular biology. Award-winning Koshland (he had just received the National Medal of Science, a major American scientific honor) then baldly states: "The article on identical twins reared apart shows that some physiological and psychological traits are inherited. . . . ," a gross overstatement. Like Bouchard, Koshland adds, "However, this does not minimize the influences of environment and motivation." Nonetheless, he continues, these new tools may reduce crime and help the mentally ill. "Although increased funding of mental health centers, stricter gun control, increased supervision of the mentally unbalanced, or higher standards for probation officers may be desirable, they are Band-Aid remedies." By linking twin studies and other research, such as noninvasive brain imaging, to the Human Genome Project, according to Koshland, the solution to these problems is clearly located in our genes. And the editorial's title, "The Rational Approach to the Irrational," is one more ideological statement that reductionist molecular genetics is the "rational," the scientific, the objectively best way of solving social problems.[48]

Six weeks later, buried in the "Letters" section, is a response from a scientist at the Human Genome Center at Lawrence Berkeley Laboratories. This supporter of the Human Genome Project objected vigorously to the distortions in that editorial, pointing to "the absurdity of suggesting that the complexity of unfortunate forces that produced a homicide in Berkeley could be fixed by the Human Genome Project. . . . To my knowledge, this is the first time that a respected scientific or biomedical journal has stated that predispositions are illnesses." The letter writer's major concern was that unrealistic promises and "slogans equating the complete genetic map to . . . the 'essence of life' " will create a backlash against the Human Genome Project.[49]

No critique of biological determinism or twin studies has been published in *Science,* and it is impossible to determine how many articles that draw attention to those critiques have been rejected. Certainly, Ruth Bleier's revealing story about *Science*'s rejection of her analysis of the flaws in sex differences research, along with the magazine's continued publication of such conceptually and methodologically flawed research, suggests that one side of certain issues is strongly favored (see chapter 5). While this preferential treatment may not be unusual for scientific publications, belying claims to objectively balanced science, certain areas of research have significant consequences for people's lives and the future of the planet. Genetic engineering is one of those areas, and the unmitigatedly enthusiastic treatment of genetic engineering in *Science* contrasts sharply with serious analyses of the complexities of the promised benefits to society.[50]

When their politics—their assumptions about power relations and the uses of molecular biology and genetics—can be uncovered through this kind of analysis, a preponderance of articles in molecular biology reflects the dominant views embraced in the leading textbooks: science is go(o)d; science is good for everyone, without distinctions; and what is good for scientists is good for the nation and the world.[51]

Even when alternative values about science-and-society are openly expressed by authors, they tend to be hidden from view in the letters section, which allows overt expressions of opinions and values. However, such views are often obscured by their headlines. For example, one letter expressed several scientists' concerns about the military use of genetically engineered pathogens (disease-causing microorganisms). The scientists make very strong position statements against the Department of Defense efforts to work with genetically engineered bio-warfare agents, particularly in aerosols, with the ostensible purpose of developing defenses against enemy use of such aerosols. The scientists express strong concern about the secrecy of military research. "It would relieve the concerns of scientists to be assured that genetic engineering will not be applied to the construction of highly dangerous biological warfare agents." The letter also refers to a discussion of the issue at a forthcoming AAAS meeting.[52]

What is the headline given to this strong and reasoned letter?: "Petition on Dugway Facility." I can imagine many other more apt headings that would not slide into journalistic sensationalism (not that *Science* is averse to sensational titles, such as "Sex and Violence" in reference to cellular functions, and "Take Back the Night," a slogan for feminist marches that protest rape and other violence against women, appropriated for a report on excess light in cities interfering with astronomical observations). Reasonable titles could be: "Genetic Engineering Proposal and the Military" or "Potential Hazards of Military Uses of Genetic Engineering." Such headings might attract readers to that letter.

These close readings of different sections of *Science* reveal a preponderance of certain kinds of political and epistemological positions. Not only are Bouchard's dubious twin studies, which support a biological determinist view of IQ, published without any reference to the serious scientific critiques of the methodology, assumptions, and interpretations of such work, but the award-winning editor of the journal highlights the work for its potential to solve crime and mental health problems. Koshland's extrapolation of this highly questionable work is just what critics of sociobiology and hereditarianism have feared: Scientifically dubious work is used toward socially oppressive ends under the guise of what is good for society. And the voices of social responsibility and responsible science are relegated to the letters bin, where they are often obscured by misleading headlines.[53]

Alternative Representations of Science and Society Relationships

Current molecular biology textbooks stand out as extreme examples of the omission of maps for students and other readers to see and strengthen links between the science described and its impact and potential consequences for people. The assumption embedded in both of the major textbooks discussed is that molecular biology is primarily recombinant DNA research and that such a molecular genetics approach to the study of life carries only positive results, with benefits to humankind. Thus, my analysis concludes that (a) although apparently silent on values issues, the textbooks give subtle but consistent messages about the relationship of this science to societal interests; (b) they assume *without qualification* that progress in molecular biology/genetics is only to the good of society, with no distinctions about differential effects on various groups or populations; and (c) when provided, the information about science-society issues is neither complete nor sufficient either to inform readers of a reasonably full range of concerns and case studies or to motivate readers to seek out more information.

Exceptions to this skewed stance on raising and addressing science-society issues are found in some textbooks in other fields of the natural sciences. A model example is a new edition of *An Introduction to Genetic Analysis,* which places the following statement as the fifth and last "key concept" listed on the title page of the introductory chapter, "Genetics and the Organism": "Genetics is of direct relevance to human affairs."[54] Indeed, the last section of that chapter addresses "genetics and human affairs," but what perspective is presented? With population biologist (and politically left) Richard Lewontin one of the coauthors, does a one-sided, gloom-and-doom politics opposite to the positivist philosophy of the molecular biology texts predominate, as charges from the radical right would lead us to believe? No, it does not. Rather, the importance of genetics to human affairs is reiterated ("Our overview of the scope of genetics would not be complete without a discussion of some of the ways in which genetics affects our everyday lives"),[55] and the authors delineate a *range* of concerns, citing specific examples of the dynamic of the science of genetics with societal issues, such as genetic engineering and biotechnology, genetic counseling, biological and environmental determination of human behavior, the impact of changing genetic varieties of food crops on costs, and the consequences of artificial fertilizers. The language clearly presents a *balance* of positive and negative ("exciting and frightening") consequences of the science of genetics:

> As with many other areas of science, new knowledge has produced new *challenges as well as solutions* to some human problems.[56] (Emphasis added.)

> Extensive planting of these [high-yield] crops around the world did provide *new food supplies, but new problems* quickly became apparent.[57] (Emphasis added.)

> But the most *exciting and frightening* applications of genetic knowledge are to the human species itself [in diagnosing hereditary diseases, genetic counseling, etc.]. . . . Our new ability to recognize genetic disease poses *an important moral dilemma.* . . . *Many other significant issues* are raised by the potential applications of genetic knowledge to human beings. . . .[58] (Emphasis added.)

The authors do not shy away from revealing that scientists have political positions and disagree among themselves: "While some scientists emphasize the promised benefits of such research, others raise disturbing questions about possible dangers."[59] This statement normalizes constructive disagreement on important issues, showing that scientists have opinions, values, commitments, political analyses, and philosophies. Information about genetics has, the chapter adds succinctly, led to information about "other new dangers as well," such as increased inherited disease and cancer from exposure to chemical food additives and other widespread chemical products and exposure to radiation from nuclear weapons fallout, nuclear reactor contamination, and some X-ray machinery.[60]

When the authors end their introductory chapter with a statement similar to the assertion by Darnell, Lodish, and Baltimore that all citizens need to understand current biology to function in today's society, readers have been given a clear idea of the many societal issues affected by genetics—and differing perspectives on these issues. This is a distinctly more balanced and more constructive representation of the interplay of science and society than that found in the molecular biology texts examined. This key point is emphasized with a drawing of the many areas of society that overlap the concerns of genetics, and a caption: "The findings of genetics have powerful impacts on many interacting areas of human behavior" (see figure 8-1.)[61] The figure gives students a clear-cut message that their understanding of genetics will be useful to them in a wide range of professions beyond scientific research and teaching, as well as in efforts for social change, thus encouraging many different students with a spectrum of political, social, intellectual, and economic concerns to continue their study of genetics.

The perspective taken on biological determinism is overtly represented in comparisons of their figures 1–4 and 1–10 (see figure 8-2), in which different models of causation (genes, environment, interaction of genes and environment, nonadditive interaction of genes, environment, and developmental noise) are built up to a complex model of the relationship of the genotype of an organism to its phenotype.[62] This clearly stated view immediately exposes the oversimplification and distortion of a biological determinist (or, alternatively, an environmental determinist) position, as well as the stance taken by many sociobiologists that the phenotype can be predicted ultimately from a percentage of genetics and a percentage of environmental input, as in Bouchard's work.

Scott Gilbert's second edition of *Developmental Biology*[63] also contains examples of efforts to balance perspectives that have often been presented in a one-sided manner. One example from the preface of this textbook shows another way that the actual interplay of science and society can be represented constructively to students:

> The text also attempts to present developmental biology as a dynamic human endeavor based on repeated observations and controlled experiments. Since the data are not independent of the people who obtained them, I have included nu-

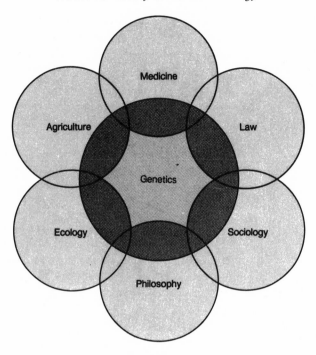

Figure 8-1. Genetics in context. From *An Introduction to Genetic Analysis 4/E,* by Suzuki, Griffiths, Miller, and Lewontin, p. 14. Copyright © 1989 by W. H. Freeman and Company. Used with permission.

merous citations so that interested individuals can read the original publications.[64]

While original citations are frequently included in many introductory textbooks, reference to the potential influence of who the researchers are on the content of science is extremely unusual.

In a third example of a constructive approach, a developmental (animal) biology textbook explicitly addresses its orientation at a complex epistemological level: "[W]e will have to recognize the symbolic nature of our system of description and abandon the notion that molecules are somehow more real than embryonic fields and that one level of description is more substantial than others."[65] The author describes his approach to developmental biology within the context of historical change in the philosophy of biology, and he explains his disagreement with the current predominant position that "such questions [of development] can be resolved in molecular terms."[66] Davenport

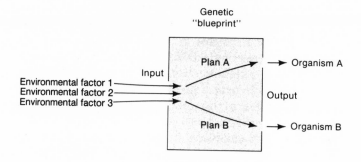

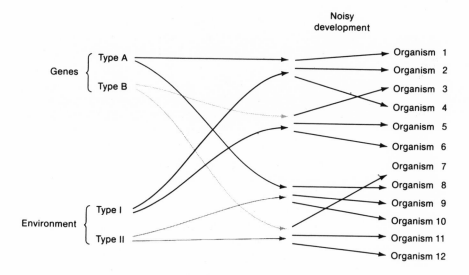

Figure 8-2. Comparison of two developmental models of input from genetics, the environment, and developmental noise. Top: "A model of determination that emphasizes the role of genes." Bottom: "A model of phenotypic determination that shows how genes, environment, and developmental noise interact to produce any given phenotype." From *An Introduction to Genetic Analysis 4/E,* by Suzuki, Griffiths, Miller, and Lewontin, pp. 6, 10. Copyright © 1989 by W. H. Freeman and Company. Used with permission.

raises issues of reductionism, objectivity, mechanistic philosophy, levels of organization, epigenesis ("machines do not arise from parts"), and a false dichotomy of experimental and descriptive studies of development.[67] Perhaps because Davenport is arguing for a position that is at odds with the prevailing ideology of hereditary concepts and molecular genetics (labeled molecular biology), he must not only present his own, but also treat it in relation to the dominant position. It is this view from the margin that requires him to delineate both the center and the margin, while someone in James D. Watson's position does not have to take the margins into account. Significantly, the 1979 edition seems to be the only version of this textbook.

Every area of science and math has developed in the context of society. A textbook that teaches about the use of computers raises an issue of "the good, the bad, and the inaccurate," making explicit in its introduction that "computers have the potential to invade our privacy or expose us to undesirable risks."[68] Part 6 of this book is entitled "Computers and Society" and addresses, along with career development and opportunities, ethical issues and social responsibilities, problems of computer security, privacy, and crime, and other social issues related to computer use. It even includes a "code of conduct and good practice" of a professional group. Clearly, this computer textbook goes well beyond a token reference to science and technology issues when it explicitly links technology to social concerns, articulating and thereby reinforcing a norm of professional responsibility in one's position and the obligation to consider the negative as well as the positive impact of computers.

When I propose that textbooks and other scientific writing should present a more inclusive and balanced view of the relationship of their field to society and the range of values affecting systems of knowledge, I am advocating awareness and information about the political content of knowledge. That is, I propose, at the very least, that it is an obligation of scientists, educators, and writers to educate students and colleagues about the actual and potential values guiding science—and the consequences of these influences on different groups in society. Making students and colleagues more aware of the differential consequences of privilege and oppression in the impact of science on society reveals the responsibilities we have as scientists, teachers, and citizens.

The current field of molecular biology creates images of the appropriate relationship of science to its social, economic, and political determinants. The evidence shows that molecular biology continues to reflect and perpetuate the belief that science is separate from society and the people producing it. This image is particularly ironic (but perhaps not accidental) during a decade in which molecular biology has become more visibly embroiled in biotechnology's industrial development, ethical issues of genetic engineering, and concerns about biohazards at the micro and macro levels of life. Paradoxically, this position assumes that, while science is value-neutral, it is also intrinsically good for society and for all people. It assumes, too, that scientists themselves know what is best for science, particularly with regard to self-regulation, defining the most important scientific problems and deciding how to best approach them. As scientists and their concerns, beliefs, and illusions become more visible, we see that who the scientists are is obviously connected with who has access to scientific knowledge and what kind of science is done, including science that supports or challenges racist and sexist beliefs. To create a truly inclusive curriculum that engages and motivates students of all backgrounds, politics, and interests, we must recognize and eliminate cultural biases that carry values favoring only a portion of the population.

Conclusion

STRENGTHENING THE BRIDGE

The danger comes when scientists allow values to introduce biases into their work that distort the results of
scientific investigations.
> —National Academy of Sciences Committee on the
> Conduct of Science, 1989[1]

[W]omen's studies brings to light the ideological nature of all structures of knowledge. Perhaps the most
important skill women's studies can pass on to students is the ability to recognize those biases [most particularly the masculine bias in curricula that once
seemed complete and impartial] where they seem most
invisible.
> —Women's Studies Task Force Report, Association of
> American Colleges, 1990[2]

As I leave for a symposium on science, technology, and gender, I realize that I am fortunate to participate in one of the ever increasing national and international efforts to strengthen the bridge between science and feminist concerns.[3] Preparing my comments on feminist critiques in cell and molecular biology, I am aware of the diversity of views I expect from the audience, particularly among scientists who are new to or even resistant to discussions of cultural bias in science and in molecular biology in particular. But I also see indications that the tide is turning toward sympathy and openness toward feminist views, as pro-women politicians, activists, educators, and scientists explode the myths of objective science and expose serious—indeed, life-threatening—biases in the degree of attention to and direction of research on, to use just one example, breast cancer.

Where are the scientists who will do the research we need to survive and to thrive? "Wanted: 675,000 Future Scientists and Engineers. A shortage of technically trained workers is looming, unless more women and minorities can be attracted to science."[4] The failure of our system of science education to create a scientifically literate populace with sufficient knowledge (not just faith) to evaluate current and future directions of science and to want to participate both as scientists and as citizens is launching numerous efforts to revamp science education.

But access to meaningful learning in science—in its real context of history, culture, and economics—must also mean access to shared values of improving the quality of life for everyone, not just those already privileged at the expense of the less powerful. As I have documented in chapter 8, despite rhetoric about the relationship of science to society, scientific communications and textbooks send the message that molecular biology is—and should be—an endeavor far distant from societal concerns. In this view, the confluence of science and sociopolitical issues is merely problematic for scientists.

Only recently has scientific commentary admitted the possibility that who a scientist is may affect which models seem right, which questions are important, which assumptions should be challenged. Even now, cultural and personal influences on science are made most obvious when the scientist is not the norm—when the scientist is a woman, a person of color, a homosexual.[5]

Science magazine, after a long record of ignoring or rejecting arguments about the relationship of sexist and racist societal beliefs to science (see chapter 5 on Ruth Bleier) while publishing research and editorials supporting biological determinist views of "difference,"[6] recently began to address the topics of women in science and minorities in science in special annual issues of the journal.[7] These highlighted sections have more images and photos of diverse women (and in the "minorities" issue, men) than ever graced a month's worth of the magazine! Most striking are the affirmatively inclusive ads: "Just what America Needs . . . Another Special Interest Group. [next page] They're men. They're women. They're black. They're white. They're Asian. They're

Hispanic. They're Argonne Scientists and Engineers. And yes, they're just what America needs"; "Many Perspectives, One Vision"; and "Excellence Knows No Gender." It remains to be seen whether this attention to fostering diversity in the scientific workforce will last.

The second (1993) issue on women benefited from hundreds of responses to the first issue, criticisms that showed *Science*'s misguided tendency to blame the victim rather than structural obstacles to women's full participation in science.[8] Most telling is the title, "Women in Science '93: Gender and Culture," that acknowledges the culture of science, but the coverage suffers from a profound misunderstanding of feminists' views about differences between women and men in relation to science, not surprising given the journal's history with sex differences claims. The most significant breakthrough, however, was *Science*'s first serious engagement with feminist critiques of science (other than the very rare book review). Despite a double-edged title, "Feminists Find Gender Everywhere in Science," and the short shrift (one and a quarter pages can only begin to correct the vast omission of the past), the article offers key insights in the brief introduction: "The point," Anne Fausto-Sterling says, "is that science can be improved by the recognition that cultural context does influence one's perspective." Elizabeth Potter adds, "It ought to be part of the scientific method . . . to look for social assumptions." Evelyn Fox Keller asserts, "My aim is to restore to science the best that science is capable of . . . it means to create a context in which everyone can make full use of the full range of human potential." The reporter's concluding statement sets a goal for science (and *Science*) to meet: "But the first, and most challenging, step toward such goals is for the feminist philosophers of science to get mainstream scientists to listen to their provocative premises."[9]

A key strategy for opening up the ears of mainstream scientists may well be seen in the billing given to the article. The heading reads "Philosophy of Science." (Other headings include "Science Education," "Women in Industry," and "Primatology," this last one the only topic addressing the content of science.) Whether strategy or putdown, the category may encourage some scientists to engage with feminist critiques of science without immediately confronting the challenges to the concept of science itself.[10] But the deeper questioning must come, once the issues are engaged.

Feminist scientists and other feminists recognize that dominant beliefs about science as a masculine pursuit are intertwined with the low participation of women in the sciences and the role of science in justifying gender inequities.[11] In her final public address before her death from cancer, Ruth Bleier said:

> I believe that the resistance of science [to feminist critiques of the content of science] is a response to a number of interrelated challenges posed by feminist criticisms. First are the challenges to positivism, a bedrock principle of Western epistemology, and to the objectivity and value neutrality that make of science, in

our society, the best if not the only route to knowledge. Moreover, unlike the analyses of traditional sociologists of scientific knowledge, feminist analyses are formulated in gendered terms. These have implications for the gendered identity, structure, and content of science, as well as implications for science's role in legitimating society's most cherished gendered beliefs and structures, namely, that hierarchical gendered social structures are based in differently gendered human natures.[12]

Cultural biases about the meaning and significance of gender and, even more broadly, of difference have done more than leave fingerprints on molecular biology. My close analysis of the best of contemporary molecular biology reveals distortions that range from inaccurate views of bacteria, fertilization, and hormones to systematically skewed definitions of the field of genetics as well as of "life" at the molecular level. Integral to these inaccuracies, the scientific community generally has insisted on preserving rather than questioning the sanctity of objectivity in regard to how and what science is done and who is allowed to do it. Despite well-intentioned proscriptions about values in science and claims of impartiality, molecular biology today presents only a partial vision of life.

As my analysis documented, beliefs in a natural complementarity of two sexes and the universality of "sex" in nature are evident in microbial physiologists superimposing labels of "male" and "female" onto bacteria, cell biologists mischaracterizing fertilization, and biochemists maintaining inaccurate designations of "male" and "female" hormones, with molecular biologists and geneticists joining the effort to extend "sex" to metabolism and developmental genetics. The paradigm of centralized control is as deeply entrenched in Western physiology as it is in molecular biology. The fields of genetics and molecular biology are both organized around nuclear heredity, with the "new" molecular biology conflated with molecular genetics. The redefinition of molecular biology to mean molecular genetics narrows our thinking about life and our ways of conceiving solutions to society's most pressing problems.

Choosing DNA as the primary controlling element in the cell fits an emerging pattern of holding male-associated entities in higher regard throughout different domains of biology: cell biology, genetics, development. This systematic skewing of the biology investigated in this study can be summarized:

Sperm as active; egg as passive

Nucleus (male) controls cytoplasm (female); gender metaphors describe power relationship

Classical genetics based on nuclear (male) heredity; neglect of cytoplasmic (female) heredity

"Sex determination" means testis (male) development

Predominant model in slime-mold aggregation is "pacemaker," centralized control, and inherent difference (masculinist)

Master molecule of life is DNA (1950s)

F plasmid (DNA) is the male signifier in *E. coli;* plasmids are essential tools
  of recombinant DNA technology
DNA is the controlling molecule of life in 1990

Biological study has elevated what is associated with males (sperm, nucleus, plasmid, DNA, Y chromosome) to a superior status, while legitimizing
dominant/subordinate relations as natural. It becomes evident that molecular
biology, cell biology, and genetics are systems of knowledge like history, literature, and political science that mirror culturally distorted beliefs in dichotomous male/female gender differences and social relations. And like other
knowledge bases, these areas of biology reinforce and further authorize such
beliefs.

My analysis also shows that the top scientists have defined molecular biology today by the techniques and assumptions of molecular genetics. Under
the terms of this definition, biology increasingly examines "life" by focusing
on a narrow conception of the "gene," while diminishing the importance of
nongenetic components and processes in the cell. Paradoxically, this view has
prevailed in spite of a clearly delineated move in biology in the last decade
away from a reductionist view of living beings and toward an explication of
complexity in gene expression and metabolism.

In tracing the effects of gender ideology on current scientific representations in this important and fascinating field of study, I have argued that recent
developments have cemented the growing domination of a reductionist and
DNA-centered form of genetics as *the* (rather than *a*) new molecular biology
of all cells, *the* most powerful and, therefore, the most correct view of life. This
hegemony can be recognized in reorganized biology curricula as well as in the
daily newspaper.

That this powerful view is only partial and incomplete must be acknowledged by scientists concerned about our scientific understanding of nature.
And to accept our responsibility as scientists and citizens, we must appreciate
how that view supports equally powerful notions of "difference" based on
dominance and subordination or superiority and inferiority, rather than variation or diversity.

We can no longer ignore the impact of cultural beliefs about gender and
difference on scientists' beliefs about who should do science and on students'
views of their own relation to science. We must consider the possibility that
the skewed ideologies embedded in the language, concepts, and guiding principles of the substance of molecular biology, as well as in the field's treatment
of science-and-society issues, may influence the participation of women and
other marginalized groups in science. Supporting this hypothesis are my interviews with women scientists, as well as years of discussions with feminist
scientists and science, math, and engineering students in which example after
example is given of students whose political and ethical concerns clash so

painfully with predominant values and practices of science that they drop out of the field.

I suggest that by reinforcing a sexist ideology of difference and empowering it with an added dimension of cellular and molecular organization of living beings—even more, by widening the perceived distance between science and society—current molecular biology unwittingly promotes the exclusion of politically aware women, people of color, and some white men from the field of molecular biology and from the new biology. The subtle ways in which these values (sexism, racism, classism, heterosexism) work, hidden in unexamined tacit assumptions and prior commitments to certain models, make them more insidious in their impact on who gets to do science and what kind of science is done. Individuals and groups suffer from the constraints of disaffection from certain areas or from science altogether. Science suffers from a decreased pool of potential talent and perspectives needed not just to "man" the labs but to enlarge our understanding and interaction with nature. Society as a whole suffers from the cultural biases of our sciences.

Overt and covert ideologies of difference, of dominant values, may encourage some science students at the expense of others, particularly those from underrepresented groups. For this reason, the study of scientific knowledge as constructed knowledge reflecting dominant values must be a necessary part of the project of enlarging participation in science, whether from a concern for equity and inclusiveness or a concern for scientific personnel needs. Exposing and critiquing all forms of gender bias in science can break the cycle, engage and empower disenfranchised students, and move scientific accuracy forward.

One view never totally predominates, and in this study I have pointed to alternative language, concepts, and paradigms within molecular biology, even within textbooks I have critiqued. These are promising examples, but deeper transformations are needed in science education, communication, and research. Training students to uncover and challenge hidden values and assumptions in the sciences must become as important as teaching them the "fundamentals" of the field; indeed, such training in critical inquiry should *be* a fundamental of the field.

Contextual critical inquiry, with an awareness of cultural ideologies, is integral to the field of women's studies and creates the bridge across the apparent gulf between feminist concerns and molecular biology. Our call for "building two-way streets"[13] means not only that women's studies programs offer science courses, integrate science and feminist critiques in their core curricula, and promote faculty development in this area, but that science faculty send their students to women's studies courses, collaborate with feminist faculty on new science/women's studies courses, and integrate science and feminist critiques into *science* curricula. Science departments and faculty must not only allow their feminist colleagues to come out of the closet about their gender-/culture-consciousness, but encourage them to do so and reward them when they contribute to integrating science-society perspectives into the science curricula. It is only when such activities become fully a part of education and research in science that science itself will approach its vision of a powerful and constructive way of knowing and understanding the world.

> The great gain [for women engaging in science] would be freedom of thought.
> —Maria Mitchell, 1871[14]

> I doubt that women as gendered beings have something new or different to contribute to science, but women as political beings do.
> —Ruth Hubbard, 1990[15]

APPENDIX A
Fields of Graduate Study in Biology

Biological Sciences

Biological and Biomedical Sciences

Anatomy

Biochemistry

Biophysics

Biotechnology

Botany and Plant Sciences

Cell and Molecular Biology

Ecology, Environmental Biology, Evolutionary Biology

Entomology

Genetics and Developmental Biology

Microbiological Sciences and Immunology (Bacteriology, Immunology, Medical Microbiology, Microbiology, Parasitology, Virology)

Neurobiology and Biopsychology

Pathology

Pharmacology and Toxicology

Physiology

Radiation Biology

Zoology (Marine Biology, Zoology)

Agricultural and Natural Resource Sciences

AGRICULTURAL SCIENCES

Agricultural Science

Agronomy and Soil Science

Food Science and Technology

Horticulture

NATURAL RESOURCE SCIENCES

Environmental Science

Fish, Game, and Wildlife Management

Forestry

Natural Resources

Range Science

Source: Peterson's Guide to Graduate Programs in the Biological and Agricultural Sciences 1989, 23d ed. Series Editor: Theresa C. Moore

APPENDIX B
Field Definitions from *Peterson's Guide*

Biochemistry

Biochemistry is a study that utilizes chemistry and other physical sciences to understand *all life processes and the products* of such processes. The study of biochemistry, therefore, is *broad* in its disciplinary application and is *broad* in the subjects and materials on which the scientist works. *The biochemist thus interacts closely with geneticists* to use chemistry to understand the mechanism of genetic transmission and the mechanism of expression of genetic information. The *biochemist also interacts with molecular biologists* to understand how nuclear material is expressed in functional cellular components. When the cellular components are studied in the context of the entire organism, there is *interaction with such biologists as physiologists and pharmacologists*. Finally, principles and methods of chemistry and other physical sciences are used to investigate *the products that result from biological systems,* such as the food supply, energy transduction, waste disposal, and toxicants. Biochemists, because of their acquaintance with chemical tools, have been largely responsible for *many recent developments in the treatment of disease and the understanding of disease processes*.

—Dr. Hector F. DeLuca, Steenbock Research Professor, Department of Biochemistry, University of Wisconsin-Madison

Cell Biology

Cell biology is the study of how animal and plant cells work. It is an *interdisciplinary, problem-oriented field, applying cell culture, biochemical, biophysical, microscopic, genetic, and immunological techniques* to answer *questions about cellular structure and function*. In the best laboratories a graduate student can expect to use *several of these different approaches* to investigate a single question. Some of the exciting problems now under investigation in cell biology are the structure, function, and biosynthesis of membranes, endocytosis and secretion, cell motility, cell-to-cell communication, biogenesis and function of organelles, packaging of DNA in chromosomes, regulation of the cell cycle, and nuclear-cytoplasmic transport. Cell biologists are attempting to *determine the mechanisms of these cellular functions at the molecular level*.

The study of cell biology encompasses and unites the fields of biochemistry, molecular biology, physiology, and structural biology. An important trend in the field is the elucidation of normal cellular functions through studies on the pathogenesis of human diseases. Graduates with *interdisciplinary training in cell biology* can qualify for positions in cell biology as well as many traditional academic departments, including *biology, biochemistry, anatomy, zoology, and botany*. In addition, there appear to be expanding opportunities for cell biologists in the *biotechnology industry*.

—Thomas D. Pollard, M.D., Professor and Director, Department of Cell Biology and Anatomy Graduate Program in Biochemistry, Cellular and Molecular Biology, Johns Hopkins Medical School

Molecular Biology

Molecular biology *evolved as a discipline in response to the need for the systematic study of the molecular and structural basis for the storage, transmission, and expression of genetic information.* With time, however, *this approach* to structure and function as it pertained to DNA and chromatin *was extended to the macromolecular systems, which provide the basis for structure and function in cell organelles other than the nucleus.* Today's molecular biology curriculum finds students examining the relation of molecular domain organization in membranes to signal transduction and immune reactions, or the relation of the molecular organization of mitochondria and chloroplasts to energy transduction and thermodynamics, or the macromolecular organization of hormone receptors in relation to gene expression. Another exciting area of research in the field of molecular biology is related to studies of gene organization and modulation of gene expression under various physiological and pathological states, including embryonic development, cell differentiation, cell transformation, and aging. Thus, molecular biology *has increasingly focused on the general study of the relationship between macromolecular organization and cellular function.*

Alongside this *broadening of the boundaries of the discipline* there has been an almost *explosive growth of the core* in the form of *DNA research and technology,* including sequencing, gene structure, and gene mapping and cataloging, so that there has been remarkable progress in our understanding of the fine structure of genes, of the sequence organization within chromosomes, and of the relations of the regulatory elements to gene expression (transcription). New developments in DNA recombination and cloning also presage rapid growth in our knowledge of gene programming and regulation. Areas such as receptor structure and function, studied from the level of gene expression to that of membrane synthesis, have gained in prominence, while the modulation of gene expression by variations in transcription and/or processing of gene sequences is now a recognized field of study.

The use of recombinant DNA technology in the construction of artificial genes with defined regulatory elements and the introduction of gene constructs into somatic cells, as well as germ-line cells (transgenic systems), has opened up a new area of research on the production of defined genotypes and the correction of genetic deficiencies and metabolic errors. In effect, the science of molecular biology, less than three decades old, has already *produced an offspring, the technology of genetic engineering,* and is rapidly *penetrating the domains of the pharmaceutical, health, and agricultural industries* in the search for agents that can predictably modulate gene expression.

The explosive growth of the field has produced a companion development, namely the high-tech biology industry based on genetic engineering, which bids fair to be *the true growth industry dominating the pharmaceutical, medical, and agricultural fields.* The industry is based on the greatly reduced time of transfer of basic research to the production of gene products. Mastery of the underlying theory and technology will give the graduate *access to jobs at all levels,* from manager to high-level technologist.

—Dr. Narayan Avadhani, Professor of Biochemistry, University of Pennsylvania

Developmental Biology

Developmental biology is a *broad, multidisciplinary* branch of biological science concerned with the establishment, maintenance, and senescence of biological systems at *all levels of organization and in all organisms*—animal, plant, and microbial. It includes the major subdivisions of reproductive biology, embryology, tissue maintenance and repair, regeneration, and aging. The field of genetics, or at least certain aspects of *genetics, is also included by many scholars as a subdivision of developmental biology* because development is directly related to regulation of the genome and gene expression. Regardless of how it is classified, no high-quality program in developmental biology can exist without a strong component of genetics. Developmental biology *shares many common interests with certain fields of medicine,* such as pathology, oncology, hematology, obstetrics-gynecology, and pediatrics, because many of the same general mechanisms that operate during normal development appear to operate to produce pathological or abnormal conditions, cancer and birth defects to name just a few. For this reason, programs in these fields sometimes include training in special aspects of developmental biology.

Because of its *multidisciplinary nature,* developmental biology has *many areas of subspecialization,* including developmental biochemistry, developmental genetics, developmental physiology, developmental morphology, *developmental cell and molecular biology,* and developmental neurobiology. Current work in all of these areas seeks to discover and describe the basic mechanisms of development and to *apply current knowledge* to prevent or correct developmental anomalies, *to increase or improve the food supply,* and to furnish new *products for industrial exploitation,* especially in the *biomedical, pharmaceutical, and food industries.*

Although there has been a lessening of demand for developmental biologists in academic institutions during the past several years, employment opportunities for university and college faculty are reasonably strong. Opportunities have increased in private and government basic research laboratories as well as in industrial applied research laboratories. There appears to be an increasing demand for professional developmental biologists in biomedical and food-production facilities.

Due to the *breadth and diversity of the field,* no graduate program can offer comprehensive coverage of all aspects of developmental biology. Each program is highly individualistic and reflects the skills and interests of its faculty. Thus the requirements and goals of each program vary somewhat. The student should take these factors into consideration when choosing a graduate program in developmental biology.

—J. Douglas Caston, Professor of Developmental Genetics and Anatomy, Case Western Reserve University

Genetics

Genetics is the study of *the inheritance of characteristics from organism to organism and the mechanisms of gene function that specify these characteristics. A wide spectrum of specializations is included* in this definition. At one end of the spectrum, molecular geneticists study how mutations—alterations in the nucleotide sequence in the DNA—alter the gene expression and hence physical characteristics; at the other end, population geneticists study how genes are conserved or lost in large groups of organisms and how new species arise. *The range of organisms so studied is striking*—bacteriophages, animal and plant viruses, bacteria, invertebrates from protozoa to fruit flies, and vertebrates from fish to human beings—although the basic concepts are quite generalized.

Current areas of activity in genetic research include the study of specific genes, using cloning techniques to determine the products and their control; the ways in which genes are put together into chromosomal arrays and the mechanisms by which chromosomes are distributed between daughter cells; inherited metabolic diseases of human beings and animals; damage to genetic material caused by environmental hazards; and the production of new, useful forms of life by genetic engineering.

Opportunities for careers in genetics are at a high point, largely because of an expanding interest in the commercial applications of gene cloning methods. In general, Ph.D.'s in genetics have tended to work in university facilities, in commercial biotechnology facilities, or in government laboratories. Well-trained students with M.S. degrees are finding positions in such areas as environmental mutagen screening programs and as technical support personnel in private and public laboratories.

—Dr. Stanley A. Zahler, Professor, and Dr. Peter J. Bruns, Professor, Section of Genetics and Development, Cornell University

Source: Peterson's Guide to Graduate Programs in the Biological and Agricultural Sciences 1989, 23d ed. Series Editor: Theresa C. Moore. (Emphasis added)

APPENDIX C
Fields and Techniques

Cytology in its early days had the central problem—what are the basic units of organisms? This problem was solved by the postulation of *the cell theory* and its subsequent elaboration and confirmation in the nineteenth century. Afterwards, the problem for cytologists (or cell biologists as they have come to be called) became the characterization of different types of cells, of organelles within cells, and of their various functions. The problem is tackled primarily with the *technique of microscopic analysis.*

The field of *genetics,* on the other hand, has as its central problem the explanation of *patterns of inheritance of characteristics.* The characteristics may be either gross phenotypic differences, such as eye color in the fly *Drosophila,* as investigated in classical genetics, or molecular differences, such as loss of enzyme activity, as investigated in modern transmission genetics. The patterns of inheritance are investigated with the *technique of artificial breeding.* The laws of segregation and independent assortment (Mendel's Laws), once their scope was known and they were well-confirmed, became part of the domain to be explained. For many of the early geneticists, though not all, the goal was to solve the central problem by the formulation of *a theory involving material units of heredity (genes) as explanatory factors.* In attempting to realize the goal, T. H. Morgan and his associates formulated *the theory of the gene of classical genetics.* Extension of the theory and techniques from *Drosophilia,* Morgan's model organism, to microorganisms marked the modern phase of the field of genetics, a phase which may be called *modern transmission genetics.*

The central problem of *biochemistry* is *the determination of a network of interactions between the molecules of cellular systems and their molecular environment;* these molecules and their interrelations are the items of the domain. As was the case with genetics in which laws became part of the domain, here too, the solution to a problem may contribute new domain items. For example, the Krebs cycle was part of the solution to the problem of determining the interactions between molecules and became, in turn, part of the domain of biochemistry; its relation to other complex pathways then posed a new problem. *Many techniques* of biochemistry are aimed at the reproduction of *in vivo* systems *in vitro,* that is, *the "test tube" simulation of the chemical reactions that occur in living things.*

The determination of *the structure and three-dimensional configuration of molecules* has become the concern of *physical chemistry.* Thus, the central problem of physical chemistry is the determination of *the interactions of all parts of a molecule relative to one another,* under varying conditions. The domain of physical chemistry is the parts of molecules and their interactions. Physical chemistry has evolved *complex techniques for the determination of the structure and conformation of molecules: x-ray diffraction, mass spectroscopy, electron microscopy, and the measurement of optical rotation.*

Source: Lindley Darden and Nancy Maull, "Interfield Theories," *Philosophy of Science* 44 (1977): 46–48. (Emphasis added)

NOTES

Preface

1. "Re-vision—the act of looking back, of seeing with fresh eyes, of entering an old text from a new critical direction—is for women more than a chapter in cultural history; it is an act of survival" (Adrienne Rich, "When We Dead Awaken: Writing as Re-Vision," in *On Lies, Secrets, and Silence* [New York: Norton, 1979], 35).

2. Women's studies is an interdisciplinary academic discipline that has celebrated an anniversary of two decades. The women's studies major was recently evaluated by the National Women's Studies Association as part of the Association of American Colleges' project to assess undergraduate liberal learning. The following statements from the report suggest the purpose and scope of women's studies as an academic field:

> [W]omen's studies both critiques existing theories and methodologies and formulates new paradigms and organizing concepts in all academic fields. It provides students with tools to uncover and analyze the ideological dynamics of their lives and become active participants in processes of social, political, and personal change. . . .
> The central organizing category of analysis in women's studies is the concept of gender, which we understand as a pervasive social construction reflecting and determining differentials of power and opportunity. From their inception, however, feminist scholarship and pedagogy also have emphasized the diversity of women's experiences and the importance of the differences among women as necessary correctives to the distortions inherent in androcentric views of human behavior, culture, and society. Women's studies therefore establishes the social construction of gender as a focal point of analysis in a complex matrix with class, race, age, ethnicity, nationality, and sexual identity as fundamental categories of social, cultural, and historical analysis. . . .
> From its position on the margin and by its willingness to identify its own ideologies, women's studies brings to light the ideological nature of all structures of knowledge—most particularly the masculine bias in curricula that once seemed complete and impartial. Perhaps the most important skill women's studies can pass on to students is the ability to recognize those biases where they seem most invisible.

Johnnella Butler et al., "Women's Studies Task Force Report," in *Reports from the Fields—Project on Liberal Learning, Study-in-Depth, and the Arts and Sciences Major,* vol. 2 (Washington, D.C.: Association of American Colleges, 1990), 208, 209, 211.

3. Jacquelyn A. Mattfeld and Carol G. Van Aken, eds., *Women and the Scientific Professions* (Cambridge: MIT Press, 1965), 22. In contrast to the interest I developed in biology and chemistry, I remember high school physics, with its ballistics and pulleys, as irrelevant to me. I was one of two females in the advanced physics class, and the

teacher made his dislike of girls, or at least of me, clear when he consistently marked wrong just enough of my correct test answers to bring my grade just below A −. When I politely pointed this out, he simply mumbled and never changed my grades. I regret his influence on me and who knows how many others.

The invisible (in that I reached this awareness only later in life) role model was my mother, who raised three children and cared for a husband and also taught young children for many years. She had majored in mathematics and minored in physics at Hunter College, graduating with honors and Phi Beta Kappa. She shared with me her love of Rachel Carson's *The Sea around Us* as we walked along the South Jersey shore, when I was still a child capable of wonder. Roselyn Solomon Spanier's love of math and ease with science and technology in everyday life (fixing toasters, appreciating rainbows) supplied me with subconscious nourishment that cannot be measured, while my father allowed his pharmacist's knowledge of and admiration for science to encourage daughter as well as son.

4. My peers from other women's colleges had a significantly higher success rate for doctorates than female cohorts who attended coeducational colleges. See Mary J. Oates and Susan Williamson, "Women's Colleges and Women Achievers," *Signs* 3 (Summer 1978): 795–806; and M. Elizabeth Tidball, "Women's Colleges and Women Achievers Revisited," *Signs* 5 (Spring 1980): 504–17. In 1967, when I graduated from Bryn Mawr College, not one biology student that I knew of in my class was admitted to medical school, while most of the students who applied to graduate schools were accepted, including those rejected from medical school. Half of the students admitted on government fellowships with me to Harvard's Biological Chemistry Department were women. In contrast, most medical schools still had a tiny quota of women, certainly less than ten percent.

5. Margaret W. Rossiter, *Women Scientists in America: Struggles and Strategies to 1940* (Baltimore: Johns Hopkins University Press, 1982). Elite women's colleges were not hotbeds of radical feminism in the early years of the women's movement. Just one example from academic literature I could have known about: sociologist Jessie Bernard's *Academic Women* (University Park, Pa.: Pennsylvania State University Press, 1964; reprint, New American Library, 1974). More politically radical and less academic was Leslie Tanner's *Voices from Women's Liberation* (New York: Mentor Books, 1970), a book I did not meet until 1979. I now know the context in which to appreciate the brilliant women scientists who were able to make their careers at women's colleges. Miss Oppenheimer is a very private person, so I do not know the particular answer to my original motivating question.

6. A key player was the chair of the biology department at Wheaton, who told me that if he suspected I would turn to the women and science topic after doing the funded laboratory research required for tenure, he would not support me for a permanent position in biology. Also, he would not allow me a two-year leave from the biology position, saying that was too long to hold my position for me. Clearly, he did not see my development in the area of women and science or the history of science as a plus in his biology department. Although I disagreed with his attitude, I never questioned his right to make those decisions. I have since learned to be more disobedient.

7. Kate Millett, *Sexual Politics* (Garden City, N.Y.: Doubleday, 1970). I wish to thank the feminists who inspired and educated me at the Bunting, including MaryAnn Amacher, Joyce Antler, Ann J. Lane, Linda Perkins, Temma Nason, Sue Standing, and Inès Talamantez.

8. Donna J. Haraway, "Investment Strategies for the Evolving Portfolio of Primate Females," in *Body/Politics: Women and the Discourse of Science,* ed. Mary Jacobus, Evelyn Fox Keller, and Sally Shuttleworth (N.Y.: Routledge, 1990), 141.

9. The original seminar produced the first version of Ruth Hubbard, Mary Sue

Henifiin, and Barbara Fried's *Biological Woman—The Convenient Myth* under the title *Women Look at Biology Looking at Women: A Collection of Feminist Critiques* (Cambridge, Mass.: Schenkman, 1979).

10. In 1980 I be fortunate to be called back to Wheaton College to administer a FIPSE-funded project, Toward a Balanced Curriculum: Integrating the New Scholarship on Women into the Curriculum; coediting a sourcebook for curriculum transformation projects; and helping women's studies grow at Wheaton. See Bonnie Spanier, Alexander Bloom, and Darlene Boroviak, *Toward a Balanced Curriculum* (Cambridge, Mass.: Schenkman, 1984); and Bonnie Spanier, "Inside an Integration Project: A Case Study of the Relationship between Balancing the Curriculum and Women's Studies," *Women's Studies International Forum* 7, no. 3 (1984): 153–59. The new Biology chair, John Kricher, gave me the great pleasure of offering a biology department course on "Women in Science: The Difference They Make," a transformational experience with wonderful students I shall never forget. In the eye-opening and challenging years spent with Wheaton's Balanced Curriculum Project, I chose women's studies as my discipline. Through a fortuitous opportunity, I was offered the women's studies program director position at Albany in 1984.

11. James Darnell, Harvey Lodish, and David Baltimore, *Molecular Cell Biology*, 1st ed. (New York: Scientific American Books, 1986), 137.

1. Molecular Biology from a Feminist Perspective

1. Warren Weaver, Letter to H. M. H. Carson, June 17, 1949, quoted in Robert Olby, *The Path to the Double Helix* (Seattle: University of Washington Press, 1974), xix.

2. James Darnell, Harvey Lodish, and David Baltimore, *Molecular Cell Biology*, 2d ed. (New York: Scientific American Books, 1990), 1.

3. Antoinette Brown Blackwell, *The Sexes throughout Nature* (New York: G.P. Putnam's Sons, 1875), quoted in *Feminists in Science and Technology Newsletter: A Publication of the Science and Technology Task Force of the National Women's Studies Association* 1, no. 1 (October 1987): 1.

4. Joan Kelly, "Did Women Have a Renaissance?" (1977), reprinted in *Women, History, and Theory: The Essays of Joan Kelly* (Chicago: University of Chicago Press, 1984), 19–50.

5. Margaret W. Rossiter, *Women Scientists in America: Struggles and Strategies to 1940* (Baltimore: Johns Hopkins University Press, 1982).

6. Patricia E. White, *Women and Minorities in Science and Engineering: An Update,* (Washington, D.C.: National Science Foundation, 1992). All statistics on participation in the sciences are affected by definitions of "scientist and engineer" that vary with disciplines included, credentialing criteria such as degree level, and data source. Social and natural science doctorates are often lumped together, obscuring the very low proportion, only about 11 percent currently, of women receiving doctorates in the physical sciences, math, and engineering.

7. Karl Mannheim's *Ideology and Utopia,* trans. Louis Wirth and Edward Shils (New York: Harcourt Brace and World, 1936) has been credited with formalizing the exclusion of science from the sociology of knowledge on the basis of the rational principles of science. Robert K. Merton's influence on the early phase of the sociology of science may have also legitimized this separation.

8. For example, Rossiter, *Women Scientists in America.*

9. For example, Carolyn Merchant, *The Death of Nature: Women, Ecology, and the Scientific Revolution* (San Francisco: Harper and Row, 1980); Evelyn Fox Keller, *Reflections on Gender and Science* (New Haven: Yale University Press, 1985); and Londa Schieb-

inger, *The Mind Has No Sex? Women in the Origins of Modern Science* (Cambridge: Harvard University Press, 1989).

10. For example, Ruth Hubbard, Mary Sue Henifin, and Barbara Fried, eds., *Biological Woman—The Convenient Myth* (Cambridge, Mass.: Schenkman, 1982); Ruth Bleier, *Science and Gender* (New York: Pergamon, 1984); Anne Fausto-Sterling, *Myths of Gender* (New York: Basic Books, 1985); Donna J. Haraway, *Primate Visions* (New York: Routledge, 1989).

11. James Darnell, Harvey Lodish, and David Baltimore, *Molecular Cell Biology*, 1st ed. (New York: Scientific American Books, 1986), viii. That book was revised and the second edition published in 1990, with that quote on page xi.

12. See chapter 2. For examples of critiques of studies of sheep, primates, humans, and even mammalian embryos: Ruth Hubbard, "Have Only Men Evolved?" in *Biological Woman*, eds. Hubbard, Henifin, and Fried, 31–33; Lila Leibowitz, " 'Universals' and Male Dominance among Primates: A Critical Examination," in *Genes and Gender, II: Pitfalls in Research on Sex and Gender*, ed. Ruth Hubbard and Marian Lowe (New York: Gordian, 1979); Donna J. Haraway, "Primatology Is Politics by Other Means" and Sarah Blaffer Hrdy, "Empathy, Polyandry, and the Myth of the Coy Female," in *Feminist Approaches to Science*, ed. Ruth Bleier (New York: Pergamon, 1986); Bleier, *Science and Gender*; Fausto-Sterling, *Myths of Gender*; Helen E. Longino and Ruth Doell, "Body, Bias, and Behavior: A Comparative Analysis of Reasoning in Two Areas of Biological Science," *Signs* 9, no. 2 (1983): 206–27; Bonnie Spanier, "The Natural Sciences: Casting a Critical Eye on 'Objectivity'," in *Toward a Balanced Curriculum*, ed. Bonnie Spanier, Alexander Bloom, and Darlene Boroviak (Cambridge, Mass.: Schenkman, 1984); and Haraway, *Primate Visions*.

13. For example, Sandra Harding and Merrill B. Hintikka, eds., *Discovering Reality: Feminist Perspectives on Epistemology, Metaphysics, Methodology, and Philosophy of Science* (Dordrecht: Reidel, 1983); Sandra Harding, *The Science Question in Feminism* (Ithaca, N.Y.: Cornell University Press, 1986); and Linda Alcoff and Elizabeth Potter, eds., *Feminist Epistemologies* (New York: Routledge, 1993).

14. The title of Sandra Harding's *Whose Science? Whose Knowledge?* (Ithaca: Cornell University Press, 1991) suggests that concern. See also Violet B. Haas and Carolyn C. Perrucci, eds., *Women in Scientific and Engineering Professions* (Ann Arbor: University of Michigan Press, 1984), especially Ruth Hubbard, "Should Professional Women Be Like Professional Men?" 205–11; and Donna J. Haraway, "Class, Race, Sex, Scientific Objects of Knowledge: A Socialist-Feminist Perspective on the Social Construction of Productive Nature and Some Political Consequences," 212–29; see also, Sue Rosser, *Female-Friendly Science* (New York: Pergamon, 1990).

15. That view is contrary to both a conservative stance that holds that women cannot be as proficient in science as men and a strictly liberal stance that women should be integrated into the scientific enterprise as it exists today. See, for example, Londa Schiebinger, "The History and Philosophy of Women in Science," *Signs* 12 (Winter 1987): 305–307.

16. Nor does it completely satisfy the ambitious goals I set for this study and its consequences:

—to open the field of molecular biology to a feminist analysis of ideologies and values embedded in the current science of molecular biology and its progeny, "the new biology," to understand how gender ideology intersects with other beliefs that guide and perhaps distort the sciences today;

—to offer constructive critiques of predominant values and ideologies for students, faculty, and researchers within the field and promote the search for alternative conceptual frameworks to provide balance;

—to advance the investigation of how feminist principles and perspectives can be useful in critical science studies of ostensibly gender-neutral subject matter;

—to advance the methodology of feminist and critical science studies with a set of general points of analysis for feminist investigations of all fields in the natural sciences, particularly to encourage other scientists to analyze gender and other ideologies in their own areas of expertise; and

—to promote investigation and discussion of the complex relationship between subject matter in a system of knowledge (which also represents a profession) and access for individuals and groups to that arena.

By extending feminist critiques and visions beyond the composition, organization, and history of the scientific professions to the substance of scientific knowledge and its epistemology, I hope to contribute to efforts toward liberation for all women and men in and out of the professions of the natural sciences.

17. She continues:

> But language, I said, is more than a game. The argument begun by feminism is not only an academic debate on logic and rhetoric—though it is that too, and necessarily, if we think of the length and influence that formal schooling has on a person's life from pre-school to secondary and/or higher education, and how it determines [a person's] social place. That argument is also a confrontation, a struggle, a political intervention in institutions and in the practices of everyday life.

Teresa de Lauretis, *Alice Doesn't: Feminism, Semiotics, Cinema* (Bloomington: Indiana University Press, 1984), 3.

18. American Association for the Advancement of Science, "Science for All Americans: Summary; Project 2061," (Washington, D.C.: AAAS, 1989). See also Bonnie Spanier, "Encountering the Biological Sciences: Ideology, Language, and Learning," in *Writing, Teaching, and Learning in the Disciplines,* ed. Anne Herrington and Charles Moran (New York: Modern Language Association of America, 1992).

19. National Academy of Sciences Committee on the Conduct of Science, *On Being a Scientist* (Washington, D.C.: National Academy Press, 1989), 6.

20. Now more than ever, Jacob Bronowski's pronouncement reflects the power of science and technology in our lives:

> There is no more threatening and no more degrading doctrine than the fancy that somehow we may shelve the responsibility for making the decisions of our society by passing it to a few scientists armored with a special magic. This is . . . the dream of H. G. Wells, in which the tall, elegant engineers rule, with perfect benevolence, a humanity which has no business except to be happy. . . . But in fact it is the picture of a slave society. . . . The world today is made, it is powered by science; and for any man [or woman] to abdicate an interest in science is to walk with open eyes toward slavery.

Jacob Bronowski, *Science and Human Values,* rev. and enl. (N.Y.: Harper and Row, 1975), 5–6.

21. Thomas Kuhn provided a simple but important example of the power of the paradigm when he described Bruner and Postman's psychological experiments with anomalous playing cards. Thomas S. Kuhn, *The Structure of Scientific Revolutions* (Chicago: University of Chicago Press, 1962), 62–65.

22. For example, Karin D. Knorr-Cetina and Michael Mulkay, eds., *Science Observed: Perspectives on the Social Study of Science* (London: Sage, 1983); and Bruno Latour and Steve Woolgar, *Laboratory Life: The Social Construction of Scientific Facts* (Beverly Hills: Sage, 1979).

This is not the place to debate the range of positions on the meaning of "the social construction of science." Readers may find inconsistencies in my position on this sub-

ject, which may be ascribed to the following: I agree with Sandra Harding that different stances with regard to empiricism, standpoint, and anti-Enlightenment
epistemologies are required based on the situation and purpose of the discussion. Stra-
tegic stances to effect change in science may differ from theoretical stances within
feminism. (See Sandra Harding, ed., *Feminism and Methodology: Social Science Issues*
[Bloomington: Indiana University Press, 1987], esp. 186.)
   23.

A list suggested by the term "sexual politics" defies termination: abortion, sterili-
zation, birth control, population policy, high technology-mediated reproductive
practices, wife beating, child abuse, family policy, definition of what counts as
a family, the sexual political economy of aging, the science and politics of diet
"disorders" and regimens, compulsory heterosexuality, heterodox sexual practices among lesbian feminists, sexual identity politics, lesbian and gay histories
and contemporary movements, rape, pornography, transsexuality, fetal and child
purchase through contract with pregnant women ("surrogacy" seems a hope-
lessly inadequate word), racist sexual exploitation, single parenting by men and
women, feminization of poverty, women's employment outside the home, unpaid
labor in the home, covertly gendered norms for professional careers, restriction to
populations of one sex in health research on non-sex-limited diseases, domestic
divisions of labor, class and race division among women, high theory in the
human sciences, technologies of representation, social research methodologies,
the ties of masculinism to militarism and especially to nuclear politics, psychoanalytic accounts of gender and culture—and on, and on.

Donna Haraway, "Investment Strategies for the Evolving Portfolio of Primate Females," in *Body/Politics: Women and the Discourse of Science,* ed. Mary Jacobus, Evelyn Fox
Keller, and Sally Shuttleworth (N.Y.: Routledge, 1990), 141.
   24. Combahee River Collective, as quoted in *Home Girls: A Black Feminist Anthology,*
ed. Barbara Smith (N.Y.: Kitchen Table, Women of Color Press, 1983), xxxi–xxxii; and
Deborah K. King, "Multiple Jeopardy, Multiple Consciousness: The Context of a Black
Feminist Ideology," *Signs* 14 (1988): 42–72.
   25. Catherine MacKinnon, "Feminism, Marxism, Method, and the State: Toward a
Feminist Jurisprudence," *Signs* 8 (1983): 636, n. 3; Peggy McIntosh, "White Privilege
and Male Privilege: A Personal Account of Coming to See Correspondences through
Work in Women's Studies," in *Race, Class, and Gender,* comp. Margaret L. Andersen and
Patricia Hill Collins (Belmont, Calif.: Wadsworth, 1992), 70–81. Such privilege can be
shared to a degree, as white heterosexual women, for example, have shared the privileges of the white men with whom they are affiliated, but the ultimate control of that
power remains in the hands of privileged white men. Documentation of the prevalence
of incest and other forms of men's physical and sexual abuse of their daughters,
nieces, wives, secretaries, maids, and other men, as well, and the loss of economic and
social status of married, middle-class white women after divorce are some indications
of the actual site of control of white male privilege.
   26. Hortense J. Spillers, "Mama's Baby, Papa's Maybe: An American Grammar
Book," *diacritics* (Summer 1987): 65–81; and Patricia Hill Collins, *Black Feminist
Thought: Knowledge, Consciousness, and the Politics of Empowerment* (Boston: Unwin
Hyman, 1990). Similarly, the particular meanings of race, class, and sexual orientation
are shaped by gender beliefs.
   27. Nancy Leys Stepan, "Race and Gender: The Role of Analogy in Science," *Isis* 77
(1986): 261–77. Reprinted in *The Anatomy of Racism,* ed. David T. Goldberg (Minneapo-
lis: University of Minnesota Press, 1990), 38–57.
   28. For a key contribution to feminist theory on the relationship of sexism to heterosexism, see Adrienne Rich, "Compulsory Heterosexuality and Lesbian Existence"

(1980). Reprinted with a foreword in *Blood, Bread, and Poetry: Selected Prose, 1979–1985* (N.Y.: Norton, 1986). For a seminal text on the historical creation of the homosexual, see Michel Foucault, *The History of Sexuality,* trans. Robert Hurley (N.Y.: Pantheon, 1978). Among many other references on the extent of homosexual activity are Neil Miller, *In Search of Gay America: Women and Men in a Time of Change* (Atlantic Monthly, 1989); and Robert E. Fay et al., "Prevalence and Patterns of Same-Gender Sexual Contact among Men," *Science* 243 (January 20, 1989): 338–48. The latter is nearly unique among *Science* articles in bringing a nonheterosexist perspective to the discussion of homosexuality.

29. In the past decade, feminist studies of science have lost some of the courageous and talented women, such as Ruth Bleier, who pioneered in this area. I particularly wish to remember the life and work of a colleague and friend, Trudy Villars, a pioneer in feminist education and psychobiology.

30. I wish to acknowledge my debt to Dr. Inès Talamantez, a colleague at the Bunting Institute of Radcliffe College in the heady days of 1978–80. Her insights and generosity helped me understand how to transform my disillusionment with the ideal of science into a responsibility for sharing the privileged information and the confidence provided by my training and experience in science.

Further, I acknowledge Sandra Harding's articulation that feminist consciousness is an achievement *(Feminism and Methodology: Social Science Issues; Science Question)*. Since feminism is an oppositional stance against the dominant ideology, values, and structures of society, feminist consciousness is an achievement that is actively sought in spite of significant pressures against it.

2. Bridging the Gulf

1. James D. Watson and John Tooze, *The DNA Story: A Documentary History of Gene Cloning* (San Francisco: W. H. Freeman, 1981), viii.

2. Donna J. Haraway, "A Manifesto for Cyborgs: Science, Technology, and Socialist Feminism in the 1980s," *Socialist Review* 15 (1985): 96. Reprinted in Donna J. Haraway, *Simians, Cyborgs, and Women: The Reinvention of Nature* (New York: Routledge, 1991).

3. National Academy of Sciences Committee on the Conduct of Science, *On Being a Scientist* (Washington, D.C.: National Academy Press, 1989), 8.

4. Aristotle, *Politica,* Book I, chapter 13; quoted in Martha Lee Osborne, ed., *Woman in Western Thought* (New York: Random House, 1979), 43.

5. For example, the phrase "rule of thumb" refers to British common law that allowed a husband to beat his wife with a stick, provided it was no thicker than his thumb. Today, the majority of U.S. states still consider it legal for a man to rape his wife.

6. While "gender" is constructed inseparably with race and class determinants, even women privileged by race, class, and ethnicity have not been considered citizens in the U.S. They did not have access to a living wage; they could not own or inherit property, vote, or implement their decisions about the uses of their own bodies.

7. A sampling of their words illustrates the claim that many of the acknowledged great minds of Western civilization reflected and promoted beliefs in a natural hierarchy of male over female, a belief that continues to infect society to this day:

Neither was the man created for the woman; but the woman for the man. (Paul)

For they who care for the rest rule—the husband the wife, the parents the children, the masters the servants; and they who are cared for obey—the women their husbands, the children their parents, the servants their masters. (Augustine)

So by such a kind of subjection woman is naturally subject to man, because in man the discretion of reason predominates. (Thomas Aquinas)

Quoted in Rosemary Agonito, *History of Ideas on Woman: A Source Book* (New York: Putnam, 1977), 72, 78, 85. See also John Locke, *Two Treatises of Government,* vol. 2, rev. ed., ed. Peter Laslett, 364 (par. 82) (New York: New American Library, 1965), reprinted in Linda J. Nicholson, *Gender and History: The Limits of Social Theory in the Age of the Family* (New York: Columbia University Press, 1986). Two other feminist analyses of the foundations of Western thought are: Susan M. Okin, *Women in Western Political Thought* (Princeton, N.J.: Princeton University Press, 1979); and Osborne, *Woman in Western Thought.*

8. She adds:

Chief among these troubling dualisms are self/other, mind/body, culture/nature, male/female, civilized/primitive, reality/appearance, whole/part, agent/resource, maker/made, active/passive, right/wrong, truth/illusion, total/partial, God/man.

Haraway, "Manifesto for Cyborgs," 96. See William Leiss, *The Domination of Nature* (Boston: Beacon, 1972), for a less explicitly feminist source of critiques of dominant/subordinate hierarchies.

9. Sandra Harding, *The Science Question in Feminism* (Ithaca: Cornell University Press, 1986), 23.

10. L. J. Jordanova, "Natural Facts: A Historical Perspective on Science and Sexuality," in *Nature, Culture, and Gender,* ed. Carol MacCormack and Marilyn Strathern (Cambridge: Cambridge University Press, 1980), 42–69; Evelyn Fox Keller, *Reflections on Gender and Science* (New Haven: Yale University Press, 1985); Carolyn Merchant, *The Death of Nature: Women, Ecology, and the Scientific Revolution* (San Francisco: Harper & Row, 1980); Susan Griffin, *Woman and Nature: The Roaring Inside Her* (New York: Harper and Row, 1978); and Elizabeth Potter, "Modeling the Gender Politics in Science," *Hypatia* 3, no. 1 (1988): 19–34.

11. Francis Bacon as quoted in Jordanova, "Natural Facts," 46.

12.

For instance, the pairs church and state, town and country also contain allusions to gender differences, and to nature and culture. Transformations between sets of dichotomies are performed all the time. Thus, man/woman is only one couple in a common matrix, reinforcing the point that it cannot be seen as isolated or autonomous.

Ibid., 43. See also Keller, *Reflections,* chapters 2 and 3.

13. The genderization of power relations has been explained by feminist object-relations advocates Dorothy Dinnerstein, Nancy Chodorow, and Jane Flax, among others. Evelyn Fox Keller ("Feminism and Science," *Signs* 7 [1982]: 589–602) draws on psychoanalytic object-relations theory as a deep explanation for why men perceive themselves as separate from and needing to control nature and women. That psychoanalysis has been a major source of support for feminist theorists is understandable, since it is the only current discourse, aside from feminism, that places sexuality and gender at the center of analysis. (I thank Mary Galvin for this insight.) However, I find economic and structural explanations of the maintenance of sexism more demonstrably effective than psychoanalytic object relations theory, particularly since the latter does not account for individual, cultural, and socioeconomic variations in women's and men's experiences nor the psychosexual heterogeneity that clearly exists in our society, suppressed and hidden though it may be.

14. Jordanova, "Natural Facts," 42–43.

15. For example, Ellen C. DuBois et al., *Feminist Scholarship: Kindling in the Groves of Academe* (Urbana: University of Illinois Press, 1987). A key contribution to the study of feminism and epistemology is, Sandra Harding and Merrill B. Hintikka, eds., *Discovering Reality: Feminist Perspectives on Epistemology, Metaphysics, Methodology, and Philosophy of Science* (Dordrecht: Reidel, 1983).

16. Margaret W. Rossiter, *Women Scientists in America: Struggles and Strategies to 1940* (Baltimore: Johns Hopkins University Press, 1982); Londa Schiebinger, *The Mind Has No Sex?: Women in the Origins of Modern Science* (Cambridge: Harvard University Press, 1989).

17. Margaret Rossiter, for example, now acknowledged with a MacArthur "genius" grant and a tenured position at Cornell, was actively discouraged by her senior colleagues and former teachers in history of science from her interest in the women cited in old editions of *American Men of Science* (*AWIS Newsletter* [1984]: 9–15). Ruth Hubbard's turn away from laboratory research to writing critiques of biology was criticized by some of her Harvard colleagues, even though a good number of her male cohorts had done the very same thing, turning their energies to political/societal concerns and letting others run their laboratories. Hubbard's protection was her tenure, while Rossiter's academic status remained remarkably unstable even, for a time, after she received the prestigious MacArthur award (Ruth Hubbard, interview, July 8, 1979).

18. Simone de Beauvoir, *The Second Sex* (New York: Vintage, 1974. Reprint of Knopf, 1953), 301; and "And Ain't I a Woman?" Sojourner Truth, in Eleanor Flexner, *Century of Struggle: The Woman's Rights Movement in the United States,* rev. ed. (Cambridge, Mass.: Belknap Press of Harvard University Press, 1975), 90–91.

19. The writings of Shulamith Firestone, Linda Gordon, Heidi Hartmann, Nancy Hartsock, Catherine MacKinnon, Adrienne Rich, and Alice Walker are examples among many. Collections of essays such as Vivian Gornick and Barbara K. Moran, eds., *Woman in Sexist Society: Studies in Power and Powerlessness* (New York: Basic Books, 1971) are noteworthy examples of a large field of scholarship that intersects with most disciplines (except the natural sciences).

20. See, for example, Carol Tavris and Carole Wade, *The Longest War: Sex Differences in Perspective,* 2d ed. (New York: Harcourt Brace Jovanovich, 1984), 221–22; Jeffrey Rubin, Frank Provenzano, and Zella Lura, "The Eye of the Beholder: Parents' Views on Sex of Newborns," *American Journal of Orthopsychiatry* 44 (1974): 512–19; and John Condry and Sandra Condry, "Sex Differences: A Study of the Eye of the Beholder," *Child Development* 47 (1976): 812–19.

Not every person who identifies as a feminist rejects beliefs in biologically determined differences between women and men. For example, Alice Rossi, feminist sociologist elected President of the American Sociological Association in 1984, shocked the feminist academic arena in 1977 by asserting that biological differences related to childbearing made women more naturally suited to parenting than men. While she suggested that compensatory education for men could change their behavior and suitability for good parenting, her position seemed no different from sociobiologists who claimed that women were better suited to mothering than to other activities, such as being scientists or sociologists. (Alice Rossi, "A Biosocial Perspective on Parenting," *Daedalus* 106 [Spring 1977]: 1–32; and Harriet E. Gross et al., "Considering 'A Biosocial Perspective on Parenting'," *Signs* 4 [Summer 1979]: 695–717.) In addition, some feminists self-identified as radical and ecofeminist believe that women are inherently superior to men, particularly in being more peaceful and loving, while men are considered more aggressive by nature. This idea is often argued on the basis of differences in testosterone levels, an argument shown to have no scientific validity.

Teresa de Lauretis plays with the charge leveled by feminists against other feminists of beliefs in essential differences in her essay in the collection, *The Essential Difference:*

Another Look at Essentialism (*differences,* vol. 1, no. 2, 1989). De Lauretis proposes that the difference that makes all the difference in the world is between being feminist or being antifeminist. ("The Essence of the Triangle or, Taking the Risk of Essentialism Seriously: Feminist Theory in Italy, the U.S., and Britain," *differences* 1 [Summer 1989]: 3–37.)

21. Both Carl Degler and the author of an article about his book use the term "nature/nurture debate." ("Revisiting the Nature vs. Nurture Debate, Historian Looks Anew at Influence of Biology on Behavior," *Chronicle of Higher Education* (May 22, 1991).

22. Thomas Bouchard et al., "Sources of Human Psychological Differences: The Minnesota Study of Twins Reared Apart," *Science* 250 (October 12, 1990): 223. See chapter 8.

23. I refer to the work of Edward O. Wilson, *On Human Nature* (New York: Bantam, 1978), 138.

24. For example, Rossiter's *Women Scientists in America* provides dramatic evidence for the barriers that have kept all but the most determined and/or privileged women out of the scientific professions.

25. See Anne Fausto-Sterling, *Myths of Gender* (New York: Basic Books, 1985), chapter 2. See also my discussion in chapter 3 regarding Benbow and Stanley's work on extraordinary math ability.

26. For example, Ruth Hubbard, "Have Only Men Evolved?" in *Biological Woman,* ed. Ruth Hubbard, Mary Sue Henifin, and Barbara Fried (Cambridge, Mass.: Schenkman, 1982), 17–46. (Also reprinted in Harding and Hintikka, *Discovering Reality,* 45–70; and Ruth Hubbard, *The Politics of Women's Biology* (New Brunswick, N.J.: Rutgers University Press, 1990.)

27. Wolfgang Wickler, *The Sexual Code: The Social Behavior of Animals and Men* (Garden City, NY: Doubleday, 1973). Quoted in Hubbard, "Have Only Men Evolved?" 31.

28. Ibid., 32.

29. "Sexual dimorphism" is defined in the *Science* issue cited as: "any differences in form regardless of whether it is manifest at the morphologic or molecular level." (*Science* 211 [March 20, 1981]: 1285.)

30. Valerius Geist, *Mountan Sheep* (Chicago: University of Chicago Press, 1971), 190, as quoted in Hubbard, "Have Only Men Evolved?" 30.

31. Ibid., 32–33. There are many examples of gender ambiguity in nature, including hermaphroditic animals and plants (many higher plants are "hermaphroditic," having both the male and female organs for producing sex cells), parthenogenic (absence of male) reproduction, and organisms that change from being "female" (egg-producing) to "male" (sperm-producing) in their life cycle.

The misrepresentation of maleness and femaleness in a study of a species with minimal sexual dimorphism—like humans—also clearly shows how the concept of ambiguous genders depends on a prior premise of a dualistic system of oppositional genders. If some other system of gender (such as a spectrum of genders without stereotyped behaviors) were the standard, our concept of gender ambiguity would not exist. See Bonnie Spanier, " 'Lessons' from 'Nature': Gender Ideology and Sexual Ambiguity in Biology," in *Body Guards: The Cultural Politics of Gender Ambiguity,* ed. Julia Epstein and Kristina Straub (New York: Routledge, 1991), 329–50.

32. Frederick Naftolin, "Understanding the Bases of Sex Differences," *Science* 211 (March 20, 1981): 1263.

33. Lynda Marie Fedigan, "Dominance and Reproductive Success in Primates," *Yearbook of Physical Anthropology* 26 (1983): 91–129; and Irwin S. Bernstein, "The Evolution of Nonhuman Primate Social Behavior," *Genetica* 73 (1987): 99–116.

34. See also Ruth Bleier, *Science and Gender* (New York: Pergamon, 1984); Sarah

Blaffer Hrdy, *The Woman That Never Evolved* (Cambridge: Harvard University Press, 1981).

35. Harding, *The Science Question,* 23. Overt expression of "metaphors of gender politics" is possible wherever sex or gender is found.

36. The Biology and Gender Study Group, "The Importance of Feminist Critique for Contemporary Cell Biology," *Hypatia* 3 (1988): 61–76. Other fields, such as organic chemistry, may also reflect cultural beliefs about the naturalness of dominant/subordinate power relations. For example, the definition of "electrophilic" displacement, from J. Stenesh, *Dictionary of Biochemistry* (New York: Wiley, 1975), is "a chemical reaction in which an electrophilic group attacks and displaces a susceptible group. . . ." This standard terminology prompted the Biology and Gender Study Group to use the consciousness-raising terms "nucleophallic and electrophallic molecules."

37. Biology and Gender Study Group, "Importance of Feminist Critique," 66, quoting G. Schatten and H. Schatten, "The Energetic Egg," *The Sciences* 23, no. 5 (1983), 29.

38. For a more detailed critique, see Emily Martin, "The Egg and the Sperm: How Science Has Constructed a Romance Based on Stereotypical Male-Female Roles," *Signs* 16 (Spring 1991): 485–501, esp. 492–98.

39. Ibid., 500, n. 70.

40. Keller, *Reflections,* part 3, 127–76. See also Potter, "Modeling the Gender Politics," for the usefulness of Mary Hesse's network model and logical coherence in feminist critiques of science.

41. Evelyn Fox Keller, "The Force of the Pacemaker Concept in Theories of Aggregation in Cellular Slime Mold," in *Reflections,* 150–57.

42. Evelyn Fox Keller, "A World of Difference," in *Reflections,* 158–76. See also Evelyn Fox Keller, *A Feeling for the Organism: The Life and Work of Barbara McClintock* (San Francisco: W. H. Freeman, 1983). Fortunately, Barbara McClintock was long-lived: the Nobel Prize is not awarded posthumously. She was an active eighty-one in 1983 when she received the award.

43. Keller, *Reflections,* 154. As she points out, "master molecule" theories preceded the identification of DNA as the genetic material; in fact, in the 1940s, proteins were the prime candidate for genes because of the known complexity of their structures compared to nucleic acids.

44. Biology and Gender Study Group, "Importance of Feminist Critique," 61–62.

45. Biological macromolecules have been classified into distinct categories—as proteins, nucleic acids, carbohydrates, and fats—on the basis of their chemical subunits, which confer particular properties. Such molecules are labelled "macro" because they are very large relative to such molecules as the oxygen we breathe; the weight of an oxygen molecule, made of two oxygen atoms, is one one-thousandth the weight of hemoglobin, a large protein composed of four polypeptide chains and a porphyrin ring containing iron.

46. For example, in 1989 Baylor University in Texas had a Department of Cell and Molecular Biology in its College of Medicine. The department's faculty includes faculty from the following departments, divisions, and institutes: Pediatrics and Microbiology/Immunology, Molecular Physiology/Biophysics, Biochemistry, Institute for Molecular Genetics, Cell Biology, Pharmacology, Division of Molecular Virology, Medicine, Pathology, Neurology, and Division of Neurosciences.

47. *Peterson's Guide to Graduate Programs in the Biological and Agricultural Sciences 1989,* 23d ed. Series ed., Theresa C. Moore (Princeton, N.J.: Peterson's Guides, 1988), 837.

48. James Darnell, Harvey Lodish, and David Baltimore, *Molecular Cell Biology,* 2d ed. (New York: Scientific American Books, 1990), xi.

49. Lindley Darden and Nancy Maull, "Interfield Theories," *Philosophy of Science* 44 (1977): 46–48.

50. For example, Darnell, Lodish, and Baltimore, *Molecular Cell Biology,* 2d ed., 11.

51. Different histories of the field reflect the perspectives of different authors. A traditional biochemistry approach is found in Joseph S. Fruton, *Molecules and Life: Historical Essays on the Interplay of Chemistry and Biology* (New York: Wiley-Interscience, 1972), while two detailed and very influential histories of molecular biology are: Horace Freeland Judson (journalist and professional writer), *The Eighth Day of Creation: Makers of the Revolution in Biology* (New York: Simon and Schuster, 1979); and Robert Olby (historian of science), *The Path to the Double Helix* (Seattle: University of Washington Press, 1974). The latter two studies reflect the lionizing of molecular biology and the major actors in the dramatic "discovery" stories. For an inquiry into contesting stories of the origins and meanings of the "catchy and flexible metaphor 'molecular biology'," see Pnina Abir-Am, "Themes, Genres and Orders of Legitimation in the Consolidation of New Scientific Disciplines: Deconstructing the Historiography of Molecular Biology," *History of Science* 23 (1985): 73–117. Jan Sapp's historical studies of genetics focus on the neglected subject of cytoplasmic inheritance (*Beyond the Gene: Cytoplasmic Inheritance and the Struggle for Authority in Genetics* [Oxford: Oxford University Press, 1987]; and *Where the Truth Lies: Franz Moewus and the Origins of Molecular Biology* [Cambridge: Cambridge University Press, 1990]).

52.

The fact is that the fashionable term "molecular biology" is unfortunate, on several grounds. Much of the research commonly held to be within this field (e.g., into the mechanisms of protein synthesis and of DNA replication) is actually quite inseparable from biochemistry; in consequence, the term is much resented by many biochemists who feel that in the eyes of the world they have no part in the currently fashionable fields which in reality are their own territory and which in a sense they were the first to explore.

John Kendrew et al., *Report of the Working Group on Molecular Biology* (London: H.M.S.O., Cmnd. 3675, 1968), 2, quoted in Fruton, *Molecules and Life,* 14.)

53. Fruton, *Molecules and Life,* vii.

54. James Watson's *Molecular Biology of the Gene* was first published in 1965 (Menlo Park, Calif.: Benjamin/Cummings) and is considered the first major textbook in the field. Subsequent editions were updated in 1970 and 1976. The fourth edition (1987) consists of two volumes and involves major revisions and additions by coauthors Nancy H. Hopkins, Jeffrey W. Roberts, Joan Argetsinger Steitz, and Alan M. Weiner.

A statement from *Molecular Cell Biology* is direct: "The modern era of molecular cell biology has been mainly concerned with how genes govern cell activity" (Darnell, Lodish, and Baltimore, 2d ed., 11.) Terminology is critical here, as "molecular biology," and "biology" are recast. Watson's title "molecular biology of the gene" made molecular genetics the focus of "molecular biology."

55. The animosity was greatest in the 1940s and 1950s. Biochemist Erwin Chargaff (who provided the nucleotide data that in a DNA molecule the quantity of adenine equals that of thymine, while cytosine equals guanine—a key piece of information toward proposing nucleotide base pairing in DNA) charged molecular biologists were "practicing biochemistry without a license" and inhibiting scientific creativity. An important founder of molecular biology, Max Delbruck, accused biochemists of "stall[ing] around in a semidescriptive manner without noticeably progressing towards a radical physical explanation." (Quoted in Scott Gilbert, "Intellectual Traditions in the Life Sciences: Molecular Biology and Biochemistry," *Perspectives in Biology and Medicine* 26 [1982]: 151–52.) Gilbert distinguishes "intellectual traditions or currents" from Kuhnian paradigms and from Foucault's singular "episteme" (151).

56. Gilbert, "Intellectual Traditions," 151–62, esp. 151.

57. Ibid., esp. 161.

58. Evelyn Fox Keller, "Making Gender Visible in the Pursuit of Nature's Secrets," in *Feminist Studies/Critical Studies,* ed. Teresa de Lauretis (Bloomington: Indiana University Press, 1986), 70, quoting from a letter from Crick to Olby in Robert Olby, "Francis Crick, DNA, and the Central Dogma," *Daedalus* (Fall 1970): 938–87. See also Evelyn Fox Keller, "Physics and the Emergence of Molecular Biology," *Journal of the History of Biology* 23 (1990): 389–409.

59. Claudio Scazzacchio, "Reflections on the Limits of Biological Reductionism," in The Dialectics of Biology Group, *Towards a Liberatory Biology,* gen. ed., Steven Rose (London: Allison and Busby Limited, 1982), 79. The new style also normalized extreme competition among scientists.

60. Richard Levins and Richard Lewontin, *The Dialectical Biologist* (Cambridge: Harvard University Press, 1985), 1–2. "Atomism" is another term used to describe this view that "the part is independent of and primordial to the whole." (Deborah R. Gordon, "Tenacious Assumptions in Western Medicine," in *Biomedicine Examined,* ed. Margaret Lock and Deborah Gordon [Dordrecht: Kluwer, 1988], 19–56, esp. 26.)

Related to the belief that the whole can be explained by understanding the parts is the use of reductionism in reference to a hierarchy of fields of study; in this case, the claim is that biology can ultimately be reduced to mathematics. See also Lynda Birke, *Women, Feminism, and Biology: The Feminist Challenge* (New York: Methuen, 1986).

61. It is argued that this is

. . . in part a result of a historical path of least resistance. Those problems that yield to the attack are pursued most vigorously, precisely because the method works there. Other problems and phenomena are left behind, walled off from understanding by the commitment to Cartesianism. The harder problems are not tackled, if for no other reason than that brilliant scientific careers are not built on persistent failure.

Levins and Lewontin, *Dialectical Biologist,* 2–3.

62. Fruton, *Molecules and Life,* 499, 501.

63. Levins and Lewontin, *Dialectical Biologist,* 2. Also, The Dialectics of Biology Group, *Liberatory Biology.*

64. Emily Martin, *The Woman in the Body: A Cultural Analysis of Reproduction* (Boston: Beacon, 1987), 12, building on Richard Lewontin, Steven Rose, and Leon J. Kamin, *Not in Our Genes: Biology, Ideology, and Human Nature* (New York: Pantheon, 1984), 11.

65. Hubbard, *Politics,* 115.

66. Fausto-Sterling, *Myths of Gender,* 48.

67. Marion Lowe, "Social Bodies: The Interaction of Culture and Women's Biology," in *Biological Woman,* 91–116; Hubbard, *Politics,* esp. chapters 8, 9, and 11; and Suzuki et al., *An Introduction to Genetic Analysis,* 4th ed. (New York: W. H. Freeman, 1989), 9, 652–53.

68. For example, Watson et al., *Molecular Biology of the Gene,* 4th ed., 25, 29.

69. One aspect of this conflated reductionism (meaning an ontological assumption, rather than just a method of investigation) involved unifying all of biological life through the apparently universal language of DNA, the linear sequence of four nucleotides. This view was symbolized by a logo for the Cold Spring Harbor Laboratory that reflected Jacques Monod's assertion that what was true for *E. coli* would also be true—through DNA—for the elephant. (Judson, *Eighth Day,* 613.) See also Abir-Am, "Themes, Genres, and Orders," on the social construction of the history of molecular biology.

70. Crick has explained that his use of "dogma" was derived from his "curious religious upbringing" and meant "an idea for which there was no reasonable evi-

dence," rather than the more commonly understood meaning of something a believer cannot doubt, a distinction that is lost on most people familiar with the concept of "the central dogma." (Judson, *Eighth Day,* 337.)

71. Recombinant DNA techniques (now called "technologies") are based, among other things, on the characterization of enzymes in bacteria that chop large pieces of DNA at specific base sequences into small, distinctive pieces that can be further analyzed and sequenced or spliced into other kinds of DNA. Another important technique uses "vectors," such as bacterial plasmids and viruses, to carry specific segments of DNA from one organism into another. See, for example, James D. Watson, John Tooze, and David T. Kurtz, *Recombinant DNA: A Short Course* (New York: Scientific American Books, 1983).

72. Some feminists (including self-identified "ecofeminists") have assumed that reductionism is inherently masculinist and that its historical opposite, organicism, is inherently profeminist. The historical study by Donna Haraway suggests the error of those theoretical assumptions and highlights the need for case studies of the intersections of various ideologies or paradigms to appreciate the contextual meaning of beliefs and the uses to which they have been put. (Donna J. Haraway, *Crystals, Fabrics, and Fields: Metaphors of Organicism in 20th-Century Developmental Biology* [New Haven: Yale University Press, 1976].) Similarly, Carolyn Merchant's historical tracings of the European metaphor of nature as female show that an organismic (or holistic) view of nature is not sufficient in itself to create or support an egalitarian gender politics.

73. See Fausto-Sterling, *Myths of Gender;* Lewontin, Rose, and Kamin, *Not in Our Genes,* esp. 19; Janet Sayers, *Biological Politics: Feminist and Anti-feminist Perspectives* (London: Tavistock, 1982), chapters 1 and 2; and portions of Sandra Harding, ed., *The "Racial" Economy of Science: Toward a Democratic Future* (Bloomington: Indiana University Press, 1993).

74. Note the distinctions that have emerged historically. Reductionism assumes that the physical parts of a whole, demarcated from its environment, account for all the properties of that entity. Applying the concept of the machine to living things, biological determinism assumes that some physical parts can be shown to cause or determine particular behaviors. Hereditarianism has become an integral part of biological determinism with the assumption that the physical determinant (such as a low forehead or a small brain) of behavior (such as criminality or inferior performance based on intelligence) is biologically inherited. (That posed a problem for sex differences in humans, since each human has a male and a female parent, but sociobiology resolves that by positing different evolutionary strategies for male and female reproduction. It still does not make sense, however.) With the development of genetics, the genes have become the physical entities that "cause" behavior.

75. Edward O. Wilson, *Sociobiology: The New Synthesis* (Cambridge: Belknap Press of Harvard University Press, 1975), 575; 555.

76. As noted earlier, Wilson's *On Human Nature* extended his arguments explicitly to assert that women would never equal men in business, politics, and science because women are inherently less aggressive than men. Another particularly egregious claim of some sociobiologists is that rape is an evolutionary adaptation as a reproductive strategy. See Fausto-Sterling, *Myths of Gender,* chapter 6.

77. For example, Ruth Hubbard and Marian Lowe, eds., *Genes and Gender, II: Pitfalls in Research on Sex and Gender* (New York: Gordian Press, 1979); Stephen Jay Gould, *The Mismeasure of Man* (New York: Norton, 1981); in addition to the Dialectics of Biology Group, Lewontin et al., Fausto-Sterling, and Birke. The interconnection between the scientific evidence and arguments and the ideology of biological determinism should not obscure the careful analyses showing invalid evidence, questionable interpretations, and unjustifiable leaps among animal models and humans. Thus, neither E. O.

Wilson's nor Alice Rossi's efforts at softening the "determinism" in biological determinism (suggesting that compensatory education of males to nurturance and females to aggression could even out the "natural" distribution of males and females in certain social roles) address the serious conceptual and methodological flaws of sociobiology (or biosociology, as Rossi attempts to distinguish her version from sociobiology).

78. Levins and Lewontin, *Dialectical Biologist,* 4.

79. Hilary Rose, "Hand, Brain, and Heart: A Feminist Epistemology for the Natural Sciences," *Signs* 9 (1983): 73–90, esp. 81–82; Ruth Hubbard, in *Liberatory Biology.*

80. For example, see Margaret C. Jacobs, *The Cultural Meaning of the Scientific Revolution* (Philadelphia: Temple University Press, 1988).

81. C. P. Benbow and J. C. Stanley, "Sex Differences in Mathematical Ability: Fact or Artifact?" *Science* 210 (December 12, 1980): 1262–64; G. B. Kolata, in "News and Views," *Science* 210 (December 12, 1980): 1235–36; and C. P. Benbow and J. C. Stanley, "Sex Differences in Mathematical Reasoning Ability: More Facts," *Science* 222 (1983): 1029–31.

3. Methodology

1. From her diary in 1871, Maria Mitchell, "Astronomer," in *Growing Up Female in America: Ten Lives,* ed. Eve Merriam (New York: Dell, 1971), 96.

2. Londa Schiebinger, *The Mind Has No Sex? Women in the Origins of Modern Science* (Cambridge: Harvard University Press, 1989), 332. I take "human" here to mean fully human in a feminist (antisexist, antiracist, antiheterosexist) sense that rejects the use of the white male as the ideal or standard of "humanness" against which "others" are measured. The feminist definition of "human" is consciously distinct from what is called the humanistic ideal, purportedly inclusive but historically gender-, race-, and class-biased.

3. Standpoint epistemology proposes that the view from the bottom up, or from the margin to the center, encompasses more information and therefore is more illuminating than the view from the top looking up or from the center looking in. The metaphor of margin to center or bottom to top assumes not only a different angle of vision on society but also recognizes that the disenfranchised are muted in society. When "they" are studied, it is not from their own perspective, so that privileged people are relatively ignorant about that sector. In contrast, predominating views and experiences (middle-class, white, male, and purportedly heterosexual) are taught and displayed everywhere. See, for example, bell hooks, *Feminist Theory From Margin to Center* (Boston: South End Press, 1984); and Nancy C. M. Hartsock, "The Feminist Standpoint: Developing the Ground for a Specifically Feminist Historical Materialism," in Sandra Harding and Merrill B. Hintikka, eds., *Discovering Reality: Feminist Perspectives on Epistemology, Metaphysics, Methodology, and Philosophy of Science* (Dordrecht: Reidel, 1983); also reprinted in Sandra Harding, ed., *Feminism and Methodology: Social Science Issues* (Bloomington: Indiana University Press, 1987), 157–80.

4. Leslie Roberts, "The Rush to Publish," *Science* 251 (January 18, 1991): 260–63.

5. For example, sperm are said to make an "existential decision" and have a "behavioral repertoire." (Bennett M. Shapiro, "The Existential Decision of a Sperm," *Cell* 49 [May 1987]: 293–94, quoted in Emily Martin, "The Egg and the Sperm: How Science Has Constructed a Romance Based on Stereotypical Male-Female Roles," *Signs* 16 [Spring 1991]: 485–501, esp. 491.)

6. James D. Watson et al., *Molecular Biology of the Gene,* 4th ed. (Menlo Park, Calif.: Benjamin/Cummings, 1987). Although the coauthors have added to and changed the book substantially, the current edition nonetheless retains much of the original glib,

sometimes arrogant, tone of Watson's original text. I will refer to this book as Watson's *Gene*. See chapter 2 for the significance of the original version of this book.

7. James Darnell, Harvey Lodish, and David Baltimore, *Molecular Cell Biology*, 1st and 2d eds. (New York: Scientific American Books, 1986 and 1990).

8.

In many respects this book is one of the boldest and most successful undertakings of its kind to date. It stands above many of its predecessors in the extent to which it presents biological facts through the experiments that made them such. It also has the attribute of pointing out where knowledge is lacking and where experimental results are equivocal. It thus gives the student an appreciation of how modern experimental cell biology has emerged, where it stands, and where it is going and invites him or her to participate in the process of unraveling the mechanisms that make cells function.

Elias Lazarides, "Modern Cell Biology," Book Reviews, *Science* 234 (December 12, 1986): 1448. I suspect that the reviewer is contrasing the attitude presented in *Molecular Cell Biology* to the arrogant and scientistic tone of Watson's *Gene*. Indeed, Darnell, Lodish, and Baltimore present a more progressive stance (actually rejecting the Central Dogma) and an approach to science education that reflects a feminist concern for empowerment, rather than mimicry.

9. Ibid.
10.

. . . it has the added attribute of relating a lot of the recent advances in molecular biology to human pathobiology.

Ibid., 1448. I do not mean to imply that the other current textbooks are not good ones. Rather, this particular one seems to reflect a more conscious intent to broaden the boundaries of "science."

11. In the preface of the first edition:

We have aimed to provide a college textbook that is no more difficult than the basic textbooks encountered by undergraduate physics and chemistry students in their respective programs of study. That there will be complaints about the scope and depth of a textbook this large seems inevitable.

Darnell, Lodish, and Baltimore, *Molecular Cell Biology*, 1st ed., viii.

The authority of one or two leading textbooks may not apply in the same way in other scientific disciplines, for example, in the social sciences, such as psychology, in which different textbooks offer competing approaches to the field (experimental, social, behavioral, Freudian, etc.).

12. Bruno Latour and Steve Woolgar, *Laboratory Life: The Social Construction of Scientific Facts* (Beverly Hills: Sage, 1979), esp. "Literary Inscription," 45–53.

The language of inscription or writing may sound strange to scientists, but the concept highlights that the products of a scientific laboratory are *representations* of parts of nature as studied with those particular instruments and techniques in that lab. As a scientist, I am tempted to add "merely representations" rather than the "real thing," but Latour and Woolgar's ethnomethodological observations of the workings and products of a scientific laboratory can make us aware of the layers of information collection, selection, and interpretation, as well as the formal and informal representations of that information, that mediate between the "real thing" (whatever that may be as an abstraction) and our human scientific creation of our representation.

My critique of the language of literary inscriptions for these representations is that such language carries a baggage of meanings from literary studies that are not only culturally specific and limited, but often limited by narrow meanings of terms within literary study, dominated as it is by techniques and ideologies of deconstruction. What I mean is that a social constructionist perspective on science can be very useful in understanding the limits of scientific knowledge, but there are risks in overlaying another questionable belief system in the process. For example, deconstruction may obfuscate political issues and lived experience by focusing solely on "the text" and abstracting all experience and communication into being "texts" to be "read," rather than animals to be observed, appreciated, interpreted.

13. W. D. Garvey and B. C. Griffith, "Scientific Communication as a Social System," *Science* 157 (1967): 1011–16; Garvey and Griffith, "Scientific Communication: Its Role in the Conduct of Research and Creation of Knowledge," *American Psychologist* 26 (1971): 349–62; and Diana Crane, *Invisible Colleges* (Chicago: University of Chicago Press, 1972).

14. Susan Wright, "Molecular Biology or Molecular Politics? The Production of Scientific Consensus on the Hazards of Recombinant DNA Technology," *Social Studies of Science* 16 (1986): 593–620.

15. Latour and Woolgar, *Laboratory Life,* 53.

16. Harding, *Feminism and Methodology.*

17. Ibid., vii. As in much of feminist thinking, a simple answer is insufficient; what is required is a reexamination and transformation of the original question in light of women's experiences. Thus, the transformed investigation becomes: let us ". . . consider what has been responsible for producing the most widely acclaimed feminist social analyses." Further, Harding states that "familiar and uncontroversial research methods have produced some of the most important new feminist analyses." Thus, the questions with which feminists are sometimes challenged ("So what's unique about feminist inquiry?" or "How can women's studies claim to be a discipline if it uses other disciplinary methods?") are revised to a constructive consideration of our own (and related) scholarship relative to traditional work. For other perspectives on feminism and methodology, see, for example, Maria Mies, "Toward a Methodology for Feminist Research," in Gloria Bowles and Renate Duelli Klein, eds., *Theories of Women's Studies* (Boston: Routledge & Kegan Paul, 1983); and Shulamit Reinharz, *Feminist Methods in Social Research* (New York: Oxford University Press, 1992).

18. Reacting against charges of "essentialism," a view that there is some female essence inherent in biological females, and wary of the long history of the misuse of biological explanations to justify social inequities and injustice, many feminists have been reluctant to include "biology" in their theorizing about how women are constructed by society. However, nonessentialist feminists are attempting to engage with a transformed sense of our "biology": to give just two examples: Janet Sayers, *Biological Politics: Feminist and Anti-Feminist Perspectives* (New York: Tavistock Publications, 1982); Ruth Hubbard, *The Politics of Women's Biology* (New Brunswick, N.J.: Rutgers University Press, 1990).

19. An analogous problem exists within feminist theorizing: the constraints of "totalizing" feminist theories, which when taken at their most extreme cannot account for either the existing diversity of experiences among women or the existence of feminist consciousness itself. Catherine MacKinnon's powerful proposals about the hegemony of male-created sexualities of dominance and subordination have disturbing corollaries as theory (rather than as consciousness-raising), one of which is that women have no self-created sense or experience of sexuality. Here, the danger of totalizing theories is that female agency seems impossible; yet women's history proves otherwise.

20. Biology and Gender Study Group, "The Importance of Feminist Critique for Contemporary Cell Biology," *Hypatia* 3 (Spring 1988): 61–62.

21. A relationship of mutuality between the seeker and the subject of inquiry brings to mind "connected knowing," a process of learning through empathy with a view different from one's own, a disciplinary stance, or the subject matter itself. See Mary Field Belenky et al., *Women's Ways of Knowing: The Development of Self, Voice, and Mind* (New York: Basic Books, 1986), esp. 112–23. Barbara McClintock's "feeling for the organism" is also similar to the concept of choosing to be on the same plane as the subject matter. See Evelyn Fox Keller, *A Feeling for the Organism: The Life and Work of Barbara McClintock* (San Francisco: W. H. Freeman, 1983).

22.

It is also relevant to note that I do not experience severe manifestations of PMS, and there is a possibility for that reason that I do not give sufficient credit to the medical model of PMS. I have, however, experienced many similar manifestations during the first three months of each of my pregnancies, so I have some sense that I know what women with PMS are talking about.

Emily Martin, *The Woman in the Body: A Cultural Analysis of Reproduction* (Boston: Beacon, 1987), 234, n. 5.

23. My experience of reclaiming my appreciation for and affiliation with science echoes that of another woman scientist, writing in one of the first collections of essays by women's liberationists in the 1960s:

The only reason that there aren't any more women scientists and technicians is because the men won't allow it. They tell us women that we are not good at mechanical skills. If we disprove their theory by learning these skills, they accuse us of being unfeminine. As a youngster I was encouraged in my scientific pursuits, yet when I tried to achieve these goals I was harassed. I had the confidence in my ability and the desire to study science. The constant pressures that I received caused me to lose both. . . . It was only after my involvement in Women's Liberation that I again found enjoyment in science.

Diane Narek, "A Woman Scientist Speaks," in Leslie B. Tanner, comp. and ed., *Voices from Women's Liberation* (New York: New American Library, 1970), 325–29.

24. Helen E. Longino and Ruth Doell, "Body, Bias, and Behavior: A Comparative Analysis of Reasoning in Two Areas of Biological Sciences," *Signs* 9 (Winter 1983): 206–27. See also, Helen E. Longino, *Science as Social Knowledge: Values and Objectivity in Scientific Inquiry* (Princeton, N.J.: Princeton University Press, 1990), esp. chapters 6 and 7.

25. Scholars have sorted the existing scholarship on feminism and science into several overlapping categories that include equity studies, epistemological concerns, and alternatives to the dominant ideology. See Sandra G. Harding, *The Science Question in Feminism* (Ithaca, N.Y.: Cornell University Press, 1986), 20–24; Carolyn Merchant, "Isis' Consciousness Raised," *Isis* 73 (1982): 398–409; Sue Rosser, "Feminist Scholarship in the Sciences: Where Are We Now and When Can We Expect a Theoretical Breakthrough?" in *Feminism and Science*, ed. Nancy Tuana (Bloomington: Indiana University Press, 1989), 3–14; and Londa Schiebinger, "The History and Philosophy of Women in Science," *Signs* 12 (Winter 1987): 306–308.

26. Body "language," illustrations, movies, and television suggest the range of communication modes not strictly included in the concept of language on which I am focusing.

27. For example, Francine Frank and Paula A. Treichler, *Language, Gender, and Professional Writing: Theoretical Approaches and Guidelines for Nonsexist Language Usage* (New York: Modern Language Association of America, 1989).

28. I adopt "pseudogeneric" to highlight the findings that the use of so-called generic male terms actually conjures up male figures, rather than females or a mixed

group. I refer to the research in which 500 junior-high-school science students were asked to make drawings based on statements describing prehistoric people. A higher proportion of students drew male-only figures of both adults and children when "generic" male terms (early man, primitive man, he, mankind) were used as compared to gender-neutral terms (early people, etc.). The author's conclusion: "Especially in science courses, where objectivity is supposedly taught by word and example, the use of such masculinely oriented terms should be abandoned" (Linda Harrison, "Cro-Magnon Woman—In Eclipse," *The Science Teacher* [April 1975]: 11).

Overtly sexist language and attitudes are rare in the formal communications of scientists to one another and to their students (for example, molecular biology textbooks avoid the generic "he" by using plural forms, "humans" or "students"). My findings parallel Rosser and Potter's studies of seventh-grade life science textbooks, used at a formative age in the generation of interest in science. Rosser and Potter do find the persistence of more subtle forms of privileging and encouraging males, such as using the male as the standard human body form in representations of the biology of humans in junior-high-school life science books and giving little information about women in science, in contrast to male-associated appeals ("Gregor Mendel is considered the father of genetics"). (Sue V. Rosser and Ellen Potter, "Sexism in Textbooks," 73–91, in Sue V. Rosser, *Female-Friendly Science: Applying Women's Studies Methods and Theories to Attract Students* [New York: Pergamon, 1990].) The book lists suggestions for equalizing access to science education for girls and women. See Rosser for work such as that of Frazier and Sadker 1973 and Kramarae 1980, documenting possible negative effects of sexism in textbooks on women and girls in the classroom.

29. A particularly egregious example is from an advanced embryology textbook still in use. Describing sexual differentiation in mammals:

The testis of the developing male fetus superimposes masculinity on a basically neutral or female state. A castrated male fetus will develop the same way as a female fetus, regardless of whether or not her ovaries are present, emphasizing not only the dominant role of maleness, but also the completely passive role of the fetal ovary.

The chapter ends:

In conclusion. . . . In all systems that we have considered, maleness means mastery, the Y-chromosome over the X, the medulla over the cortex, androgen over oestrogen. So physiologically speaking, there is not justification for believing in the equality of the sexes; *vive la différence!*

C. R. Austin and R. V. Short, eds., *Reproduction in Mammals.* Book 2: *Embryonic and Fetal Development* (Cambridge: The University Press, 1972), 56, 70. My thanks to Meredith Gould-Somero for bringing this to my attention many years ago.

30. Evelyn Fox Keller, *Reflections on Gender and Science* (New Haven: Yale University Press, 1985).

31. Sharon Traweek, *Beamtimes and Lifetimes: The World of High Energy Physicists* (Cambridge: Harvard University Press, 1988), chapter 3.

32. Roberta Hall and Bernice R. Sandler, *The Classroom Climate: A Chilly One for Women?* (Washington, D.C.: Project on the Status and Education of Women, Association of American Colleges, 1982). Anyone who doubts that sexist and other inappropriate differential treatment of women in colleges and universities by their teachers is a problem should consult that publication, the title of which is now used as a general reference to more covert forms of discrimination and exclusion. Sexual harassment occurs in the laboratory and offices as well as in the classroom. One prominent biomedical researcher is known to put his arm around any woman in his lab and fondle

her breasts; only one woman is known to have stopped his harassment (but only of her) by grabbing his crotch. Several years ago, I passed a large poster hanging on the entry door to a research lab in Harvard's Biochemistry and Molecular Biology Department. It was blatantly offensive to women, expressing the sentiment that women should enter at their own risk. I now regret that I was too appalled and embarrassed to stop, at the very least, and write down what it said and whose lab it was.

33. Or it may not. Physics and astronomy have black holes, chemistry has passive and active atoms and matter and nucleophilic attack.

34. Keller, *Reflections;* Carolyn Merchant, *The Death of Nature: Women, Ecology, and the Scientific Revolution* (New York: Harper and Row, 1980).

35. Keller, *Reflections,* 134. Furthermore, biological determinist claims about "human nature" mostly justify sexist and racist arrangements. The significance of such beliefs is suggested by the view of African American Clarence Thomas, then federal appeals judge and Supreme Court nominee to replace Thurgood Marshall, in his praise of an anti-abortion essay by the Heritage Foundation's Lewis Lehrman. Thomas's speech to the Heritage Foundation noted that the essay was "a splendid example of applying natural law," specifying that "human nature provides the key to how men [*sic*] ought to live their lives." ("Thomas Endorsed Anti-abortion Stance," *Times-Union* [Albany] July 3, 1991, A1.)

36.

> To McClintock, nature is characterized by an a priori complexity that vastly exceeds the capacities of the human imagination. Her recurrent remark, "Anything you can think of you will find," is a statement about the capacities not of mind but of nature. . . . Organisms have a life and an order of their own that scientists can only begin to fathom. "Misrepresented, not appreciated. . . . [they] are beyond our wildest expectations. . . . They do everything we [can think of], they do it better, more efficiently, more marvelously." In comparison with the ingenuity of nature, our scientific intelligence seems pallid. It follows as a matter of course that "trying to make everything fit into set dogma won't work. . . . There's no such thing as a central dogma into which everything will fit."

Keller, *Reflections,* 162.

37. The term *paradigm* has multiple meanings, as its popularizer Thomas Kuhn discusses in the postscript to the second edition of *The Structure of Scientific Revolutions* (Chicago: University of Chicago Press, 1970), 175. Kuhn's update settles on the following twofold meaning; on the one hand, a paradigm can be:

> the entire constellation of beliefs, values, techniques, and so on shared by the members of a given community. On the other hand, it denotes one sort of element in that constellation, the concrete puzzle-solutions which, employed as models or examples, can replace explicit rules as a basis for the solution of the remaining puzzles of normal science.

Thus, paradigms, large and small, function as implicit rules.

38. Longino and Doell, "Body, Bias, and Behavior," 224.

39. Anne Fausto-Sterling, "Society Writes Biology/Biology Constructs Gender," *Daedalus* (Fall 1987): 61–76, esp. 68–69.

40. Sarah Blaffer Hrdy, *The Woman That Never Evolved* (Cambridge: Harvard University Press, 1980); and Hrdy, "Empathy, Polyandry, and the Myth of the Coy Female," in Ruth Bleier, ed., *Feminist Approaches to Science* (New York: Pergamon, 1986). See Donna J. Haraway, *Primate Visions: Gender, Race, and Nature in the World of Modern Science* (New York: Routledge, 1989), chapter 15, for an analysis of Hrdy's approach to primate studies and sociobiology.

41. See Harding, *The Science Question,* especially chapter 5; and Longino, *Science as Social Knowledge.*

42. Anne Fausto-Sterling, *Myths of Gender* (New York: Basic Books, 1985).

43. C. P. Benbow and J. C. Stanley, "Sex Differences in Mathematical Ability: Fact or Artifact?" *Science* 210 (December 12, 1980): 1262–64. See G. B. Kolata, in "News and Views," 1235–36; C. P. Benbow and J. C. Stanley, "Sex Differences in Mathematical Reasoning Ability: More Facts," *Science* 222 (1983): 1029–31; Lynn H. Fox, "Sex Differences among the Mathematically Precocious," *Science* 224 (1984): 1291–93; and Lynn H. Fox and Sanford J. Cohn, "Sex Differences in the Development of Precocious Mathematical Talent," in *Women and the Mathematical Mystique,* ed. Lynn H. Fox, Linda Brody, and Dianne Tobin (Baltimore: Johns Hopkins University Press, 1980). See also Fausto-Sterling, *Myths of Gender,* 53–59. Furthermore, recent studies document the significant impact of Benbow and Stanley's claims (which were highlighted in newspapers, *Time* ["The Gender Factor in Math," Dec. 15, 1980, p. 57], and *Newsweek* ["Do Males Have a Math Gene?" Dec. 15, 1980, p. 73]) on parental attitudes toward girls' and boys' math abilities (Jacquelynne S. Eccles and Janis E. Jacobs, "Social Forces Shape Math Attitudes and Performance," *Signs* 11 [Winter 1986]: 367–80).

44. See Harding, *The Science Question;* and Harding and Hintikka, *Discovering Reality.*

45. Keller, *A Feeling for the Organism;* and James D. Watson and John Tooze, *The DNA Story: A Documentary History of Gene Cloning* (San Francisco: W. H. Freeman, 1981), viii. See epigraph in chapter 2 and analysis in chapter 8.

46. Susan Bordo, "The Cartesian Masculinization of Thought," *Signs* 11 (1986): 439–56, esp. 452–54; Susan Griffin, *Pornography and Silence: Culture's Revenge against Nature* (New York: Harper and Row, 1981); and Susan Griffin, *Woman and Nature: The Roaring inside Her* (New York: Harper and Row, 1978).

47. For example, in what ways can scientists themselves participate in and benefit from the perspectives and tools of the social studies of science? Steve Woolgar, ed., *Knowledge and Reflexivity: New Frontiers in the Sociology of Knowledge* (London: Sage, 1988).

48. Autumn Stanley, presentation at National Women's Studies Association Meeting, 1984.

49. Darnell, Lodish, and Baltimore, *Molecular Cell Biology,* 2d ed., 10–11.

50. The minor error is unfortunate but not surprising, since Wilson is usually credited with the discovery without mention of Stevens. A biologist of Morgan's stature, Wilson came to similar conclusions about sex determination by chromosomes, although Brush argues that Nettie Stevens probably came to them first and with evidence more clearly showing the dominant/recessive relationship of the X- to the Y-chromosome because of the difference in size. (James D. Watson, et al., *Molecular Biology of the Gene,* 4th ed. [Menlo Park, Calif.: Benjamin/Cummings, 1987], 12; Steven G. Brush, "Nettie M. Stevens and the Discovery of Sex Determination by Chromosomes," *Isis* 69 [June 1978]: 163–72.)

51. Darnell, Lodish, and Baltimore, *Molecular Cell Biology* 2d ed., 11.

52. James D. Watson, *The Double Helix: A Personal Account of the Discovery of the Structure of DNA* (New York: Atheneum, 1968). Watson's molecular biology textbook also credits Franklin alongside Wilkins and reproduces the key X-ray crystallograph taken by Franklin and Gosling. Wilkins shared the Nobel Prize with Watson and Crick in 1962; Franklin had died by then, but were she alive it is probable that she would not have been recognized. Sir Lawrence Bragg's foreword in *The Double Helix* supports the devaluing of Franklin's work by pointing with pride to Wilkins's "long, patient investigation," without making any reference to Rosalind Franklin (viii).

53. Anne Sayre, *Rosalind Franklin and DNA* (New York: Norton, 1975); Ruth Hubbard, "Reflections on the Story of the Double Helix," *Women's Studies International Quarterly* 2 (1979): 1–13; and more recent versions: Hubbard, "The Story of DNA," in

Structures of Matter and Patterns in Science, Inspired by the Work and Life of Dorothy Wrinch, 1894–1976, ed. Marjorie Senechal (Cambridge, Mass.: Schenkman, 1980), 117–38; and Hubbard, "The Double Helix: A Study of Science in Context," in *The Politics of Women's Biology,* 48–66.

Kenneth R. Manning's *Black Apollo of Science: The Life of Ernest Everett Just* (New York: Oxford University Press, 1983) provides insights into devastating constraints on the few African American scientists attempting to enter mainstream research in the first half of the twentieth century. The marginalization of Just's female counterpart, Roger Arliner Young, was even more extreme and disheartening. I thank Evelynn Hammonds for her insights on Manning's treatment of Young. See also Kenneth R. Manning, "Roger Arliner Young, Scientist," *Sage: A Scholarly Journal on Black Women* 6, no. 2 (Fall 1989): 3–7.

54. Darnell, Lodish, and Baltimore, 2d ed., 8. Although this textbook clearly features McClintock, ironically, Harriet Creighton, her female colleague in the earlier work on chromosome breakage and rejoining, is overlooked. The Watson et al. *Gene* textbook refers correctly to both women (14).

55. Quote is from Roger Lewin, "A Naturalist of the Genome," *Science* 222 (October 28, 1983): 402, on McClintock's Nobel Prize for Physiology and Medicine.

In the twentieth century, the study of plants has been viewed as less central to biology. One factor that influenced the relative obscurity of Barbara McClintock's work on "jumping genes," or movable genetic elements, in the 1940s and 1950s was that the work was on corn. Not until the same phenomenon was detected in yeast and fruit flies did it (and McClintock) get more recognition. The genetic workings of plants differ in many ways from those of animals and remain only partially understood. For example, a plant species may have multiple copies of the DNA in the chromosomes without being very different from a closely related species with but one copy. A biology and genetics based more on the study of plants might be oriented differently from our current biology and genetics. I leave it to my feminist colleagues with expertise in plants to explore this speculative proposal.

For more information about Barbara McClintock's contributions to science and society, see Keller, *A Feeling for the Organism,* and Nina Federoff and David Botstein, eds., *The Dynamic Genome: Barbara McClintock's Ideas in the Century of Genetics* (Cold Spring Harbor, N.Y.: Cold Spring Harbor Laboratory Press, 1992).

56. *Peterson's Guide to Graduate Programs in the Biological and Agricultural Sciences 1989,* 23d ed. series ed., Theresa C. Moore (Princeton, N.J.: Peterson's Guides, 1988), 2. Note that the 1991 (25th) edition only states: "There are now more foreign students, more women, and more students who are members of minority groups" (2).

57. This is based on selected 1993 issues of *Cell, Journal of Biological Chemistry, Molecular and Cell Biology,* and *Science.*

58. Stacey Young, "Gender Bias in Math Texts," unpublished paper, University at Albany, S.U.N.Y., 1984.

59. See, for example, Sherry Turkle and Seymour Papert, "Epistemological Pluralism: Styles and Voices within the Computer Culture," *Signs* 16 (Autumn 1990): 128–57. See also, in the growing area of ethnomathematics: Marilyn Frankenstein, *Relearning Mathematics: A Different Third R—Radical Math(s),* vol. 1 (London: Free Association Books, 1989); G. G. Joseph, "Foundations of Eurocentrism in Mathematics," *Race and Class* 28, no. 3 (1987): 13–28; and Paulus Gerdes, "On Culture, Geometrical Thinking, and Mathematics Education," *Educational Studies in Mathematics* 19 (1988): 137–62.

60. Informal communications and letters circulated to members of the AMS (American Mathematical Society), 1986–87. I thank Dusa McDuff for sharing these: *Newsletter of Association for Women in Mathematics,* May–June, 1986; July–August, 1987. As

reported in *Science,* 57% of the voting members agreed to "lend no support to the Star Wars program" and to "make no efforts to obtain funding for Star Wars research or to mediate between agencies granting Star Wars funds and people seeking these funds." Thirty-three percent of the voters were against the proposal and 10 percent abstained. Other motions for nonmilitary support for mathematics research and basic research received overwhelming approval. C.N., "Mathematicians Say No to SDI Funding," *Science* 240 [April 8, 1988]: 140.)

61. Faye Flam, ed., "Math Societies Cancel Denver Meeting," *Science* 259 (February 12, 1993): 898. The small news report was unusual for its attention to homosexuality and civil rights.

4. Sex and the Single Cell

1, James D. Watson et al., *Molecular Biology of the Gene,* 4th ed. (Menlo Park, Calif.: Benjamin/Cummings, 1987), 191.

2. "Does Ideology Stop at the Laboratory Door? A Debate on Science and the Real World," *New York Times,* October 22, 1989.

3. Watson et al., *Gene* 190, 191.

4. Ibid., 132. Experiments in mapping the genome of the bacteria used agitation with a Waring blender to separate the cells connected during genetic transfer; while the textbook adds that these experiments were called "the blender experiments," they are referred to informally as "coitus interruptus."

5. Heterosexist assumptions are as much at the core of the problem of biological sex differences for biologists studying single-celled organisms such as chlamydomonas as they are for those studying Bighorn sheep (see chapter 2). If "male" and "female" represent inherent difference, how could genetically identical cells "copulate"? See Jan Sapp, *Where the Truth Lies: Franz Moewus and the Origins of Molecular Biology* (Cambridge: Cambridge University Press, 1990), chapter 3, "Sex and the simple organism," 56–83, esp. 70–73. For more on gender ambiguity in nature and the rejection of models that do not adhere to binary sex, see Bonnie Spanier, " 'Lessons' from 'Nature': Gender Ideology and Sexual Ambiguity in Biology," in *Body Guards: The Cultural Politics of Gender Ambiguity,* ed. Julia Epstein and Kristina Straub (New York: Routledge, 1991).

6. The Biology and Gender Study Group, "The Importance of Feminist Critique for Contemporary Cell Biology," *Hypatia* 3 (Spring 1988): 71; T. M. Sonneborn, "Sexuality in Unicellular Organisms," in *Protozoa in Biological Research,* ed. Gary N. Calkins and Francis M. Summers (New York: Columbia University Press, 1941). See also Sapp, *Where the Truth Lies.*

7. James Darnell, Harvey Lodish, and David Baltimore, *Molecular Cell Biology,* 1st ed. (New York: Scientific American Books, 1986), 137.

8. Sandra M. Gilbert and Susan Gubar, *The Madwoman in the Attic* (New Haven: Yale University Press, 1979), 3, 7.

9. Ibid., 4.

10. Feminist/women scientists have a job similar to that of feminist/women writers: not only to wrest the plasmid from the exclusively male hand (doing what?), but also to transform the meaning of the tool as well as the functions to which people put it. That significant concern is beyond the scope and intent of this project.

11. Another example is the comment in Figure 7-17 in Watson et al., *Gene* 192: "The size of the F DNA is exaggerated for clarity."

12. Darnell, Lodish, and Baltimore, *Molecular Cell Biology,* 1st ed., 992–94. One of the major changes in the second edition was to integrate treatment of development and cellular differentiation throughout the text, rather than segregating it in a chapter of its own. As a result, it was difficult to determine if the discussion of fertilization was

deleted, as it seems to be, in the second edition. The index did not list "fertilization" or "egg." It did list "sperm" with the qualifier, "acrosome reaction," and the text, although concerned only with the acrosome reaction during fertilization, was not very different from the original description of fertilization (2d ed., 1990, pp. 886–88 with figures). That is, the sperm and its components (proteins, membrane, enzymes) were given the active role. The egg, it seems, has been nearly eliminated from the content of the new edition.

13. See chapter 2. Biology and Gender Study Group, "Importance of Feminist Critique," esp. 66; and Emily Martin, "The Egg and the Sperm," *Signs* 16 (Spring 1991): 485–501.

14. Darnell, Lodish, and Baltimore, *Molecular Cell Biology,* 1st ed., 994.

15. Ibid., 840–41, 994.

16. Darnell, Lodish, and Baltimore, *Molecular Cell Biology,* 2d ed., 888.

17. See Biology and Gender Study Group, "Importance of Feminist Critique," esp. 62–68; Martin, "Egg and Sperm."

18. Emily Martin, *The Woman in the Body: A Cultural Analysis of Reproduction* (Boston: Beacon, 1987), 48.

19. See Biology and Gender Study Group, "Importance of Feminist Critique," 61–62.

20. Darnell, Lodish, and Baltimore, *Molecular Cell Biology,* 2d ed., 1–11.

21. Ibid., 7.

22. For a recent analysis of the history of national and cultural differences in perspective on nuclear and cytoplasmic genetics, see Jan Sapp, *Beyond the Gene: Cytoplasmic Inheritance and the Struggle for Authority in Genetics* (New York: Oxford University Press, 1987).

23. Watson et al., *Gene* 22.

24. Darnell, Lodish, and Baltimore, *Molecular Cell Biology,* 2d ed., 437–44.

25. Joseph Palca, "The Other Human Genome," *Science* 249 (September 7, 1990): 1104.

26. The rest of that paragraph:

For example, mitochondrial genes are inherited only from the mother. And when a cell divides, mitochondria are randomly assigned to the daughter cells. Since there are many mitochondria per cell (and each mitochondrion contains between four and ten copies of the mtDNA genome), a mutation may be present in the mtDNA somewhere in the cell, but it can be far outnumbered by the normal, nonmutated copies that will also be present. Therefore, mtDNA-linked changes in an organism may not show up until the mutation spreads to a sufficient number of mitochondria.

Ibid.

27. Sapp, *Where the Truth Lies,* 1.

28. The Fathers' Rights Association of New York State, Inc. has criticized the prochoice stance of, for example, the National Organization for Women for not addressing "the rights of fathers or the lives of unborn children" indicating "a desire to deprive men of reproductive freedom and to disregard the rights of unborn children" (from a letter by Peter G. Sokaris, Albany, N.Y., to the editor of the *Times-Union* (Albany), November 17, 1988).

29. James Nelson at the Hastings Center for Medical Ethics Studies has pointed out that, by devaluing the woman's role in gestation, women are made "equal" with men. "Motherhood becomes strictly the contribution of a gamete," he said, "in her case an ovum rather than a sperm." (Seth Mydans, "Science and the Courts Take a New Look at Motherhood," *New York Times,* November 4, 1990.)

30. For a thought-provoking analysis of genetics, see Ruth Hubbard, "Genes as Causes," in *The Politics of Women's Biology* (New Brunswick, N.J.: Rutgers University Press, 1990), 70–86.

5. Sex and Molecules

1. Graeme Mardon et al., "Duplication, Deletion, and Polymorphism in the Sex-Determining Region of the Mouse Y Chromosome," *Science* 243 (January 6, 1989): 78.
2. Ibid.
3. Eva M. Eicher and Linda L. Washburn, "Genetic Control of Primary Sex Determination in Mice," *Annual Review of Genetics* 20 (1986): 327–60, as quoted in Anne Fausto-Sterling, "Life in the XY Corral," *Women's Studies International Forum* 12 (1989): 329.
4. "Sister chromatids" (identical copies of chromosomes arising during duplication) and "daughter cells" (cells resulting from cell division in mitosis) are common terms from cell biology. The choice of these female designations may be based on the association of reproduction with the female, but it creates unnecessarily gendered language. In contrast, "maternal RNA" is faithful to the meaning "coming from the mother," as it refers to RNA in the cytoplasm of a fertilized egg, originating in the egg prior to fertilization.
5. Margaret L. Andersen, *Thinking about Women: Sociological and Feminist Perspectives* (New York: Macmillan, 1983), 31–33.
6. The vaginal smear test (in mice and rats) was the biological assay chosen in 1932 to detect and measure "female" hormones (actually, estrogen).

> The acceptance of the vaginal smear test as the standard test for female sex hormones brought a definite change in the way the test was used. Where it had formerly been an instrument for investigating the function of ovarian preparations, it was now a tool for attaching the label "female" to chemical substances. Using this test as the decisive criterion, researchers found that an unexpectedly large number of substances could be labelled as "female." According to the original concept of female sex hormones, these substances were thought to originate exclusively from female organs: the ovaries. Applying the vaginal smear test, scientists had to conclude that female substances were present in a wide variety of organs, did not originate only in animals, and were not even restricted to the female of the species. . . . Now substances from such varied sources as yeast, willow buds, potatoes, sugar beets, rice, placental tissue, the body fluids of males (e.g., blood, urine, and bile), and even testicular tissue could be labeled "female."

Nelly Oudshoorn, "On Measuring Sex Hormones: The Role of Biological Assays in Sexualizing Chemical Substances," *Bulletin of the History of Medicine* 64 (1990): 243–61; quote, 251.
7. Herbert Evans, 1939 review of "Endocrine Glands: Gonads, Pituitary, and Adrenals," as quoted in Diana Long Hall, "Biology, Sex Hormones, and Sexism," in *Women and Philosophy: Toward a Theory of Liberation,* ed. Carol C. Gould and Marx W. Wartofsky (New York: Putnam, 1976), 81–96; quote, 91. For a more recent study, see Nelly Oudshoorn, "Endocrinologists and the Conceptualization of Sex," *Journal of the History of Biology* 23, no. 2 (1990): 163–87.
8. Hall, "Biology, Sex Hormones, and Sexism," 89.
9. Frederick Naftolin, "Understanding the Bases of Sex Differences," *Science* 211 (March 20, 1981): 1264. The author was then professor and chair of the Department of Obstetrics and Gynecology at the Yale School of Medicine.
10. Londa Schiebinger, "Skeletons in the Closet: The First Illustrations of the Fe-

male Skeleton in Eighteenth-Century Anatomy," *Representations* 14 (Spring 1986): 42–82, esp. 48–49.

11. Ibid., 43.

12. Ibid., 56.

13. Mardon et al., "Duplication, Deletion," 78.

14. Anne Fausto-Sterling, "Society Writes Biology/Biology Constructs Gender," *Daedalus* (Fall 1987): 61–76; quote, 64–65.

15. Bruce M. Carlson, *Patten's Foundations of Embryology,* 4th ed. (New York: McGraw-Hill, 1981), 459, as quoted in Fausto-Sterling, "Society Writes Biology," 65.

16. Fausto-Sterling, "Life in the XY Corral."

17. David C. Page et al., "The Sex-Determining Region of the Human Y-Chromosome Encodes a Finger Protein," *Cell* 51 (1987): 1091–1104; Georges Guellan et al., "Human XX males with Y single-copy DNA fragments," *Nature* 307 (1984): 172–73.

18. Studies of general attitudes about what constitutes male and female in humans have shown that the presence or absence of a penis is the most significant cultural determinant. (Suzanne J. Kessler and Wendy McKenna, *Gender: An Ethnomethodological Approach* [1978]; reprint, Chicago: University of Chicago Press, 1985.) More recently, Kessler has reported the significance of the potential for a non-normal-sized penis in the medical adjudication of intersexed infants, regardless of the chromosome complement of XY, the presence of testes, and the production of normal quantities of androgen. That is, doctors recommend surgery to make an XY individual with a functional but small penis (a micropenis) into a "female." (Suzanne J. Kessler, "The Medical Construction of Gender: Case Management of Intersexed Infants," *Signs* 16 [Autumn 1990]: 3–26, esp. 21–23).

19. Eicher and Washburn, "Genetic Control," 328–29, as quoted in Fausto-Sterling, "Life in the XY Corral," 329.

20. Mardon et al., "Duplication, Deletion," 79.

21. Claude M. Nagamine et al., "Chromosome Mapping and Expression of a Putative Testis-Determining Gene in Mouse," *Science* 243 (Jan 6, 1989): 80–83; quote, 80.

22. I cite primarily from an issue of *Science* devoted to "Sexual Dimorphism," ed. Frederick Naftolin and Eleanore Butz, *Science* 211 (March 20, 1981). While that issue was published a number of years ago, current articles reflect no significant change in assumptions and methodology; for example, articles published in *Science* 243 (January 6, 1989) and cited above. *Science* is typical of scientific journals and other publications in using the dominant paradigm of sex determination. See a detailed critique in Fausto-Sterling, "Society Writes Biology."

23. See Kessler, for example, for physical ambiguities; and Julia Epstein and Kristina Straub, eds., *Body Guards: The Cultural Politics of Gender Ambiguity* (New York: Routledge, 1991), for cultural and historical examples.

24. Anke A. Ehrhardt and Heino Meyer-Bahlburg, "Effects of Prenatal Sex Hormones on Gender-Related Behavior," *Science* 211 (March 20, 1981): 1313. The original Money and Ehrhardt research on "masculinized" girls, exposed prenatally to androgens, and subsequent claims about that work have been comprehensively critiqued; for example, Anne Fausto-Sterling, *Myths of Gender: Biological Theories about Women and Men* (New York: Basic Books, 1985), chapter 5: "Hormones and Aggression: An Explanation of Power," esp. 133–41; and Ruth Bleier, *Science and Gender* (New York: Pergamon Press, 1984), esp. 97–103.

25. Ehrhardt and Meyer-Bahlburg, "Effects of Prenatal," 1316. For an in-depth analysis, see Helen E. Longino, *Science as Social Knowledge: Values and Objectivity in Scientific Inquiry* (Princeton, N.J.: Princeton University Press, 1990).

26. Anke A. Ehrhardt et al., "Sexual Orientation after Prenatal Exposure to Exogenous Estrogen," *Archives of Sexual Behavior* 14 (1985): 57–77.

27. Bruce McEwen, "Neural Gonadal Steroid Actions," *Science* 211 (March 20, 1981): 1310.

28. The verb "obscure" is used this way in the introductory essay in this issue on sexual dimorphism (Naftolin, "Understanding the Bases," 1264).

29. See Fausto-Sterling, *Myths of Gender;* and Bleier, *Science and Gender.*

30. Ehrhardt and Meyer-Bahlburg, "Effects of Prenatal," 1312.

31. C. DeLacoste-Utamsing and R. L. Holloway, "Sexual Dimorphism in the Human Corpus Callosum," *Science* 216 (1982): 1431–32. This "misadventure" in science was published posthumously. See Ruth Bleier, "A Decade of Feminist Critiques in the Natural Sciences," *Signs* 14, no. 1 (Autumn 1988): 186–95, esp. 191–93.

32. Bleier, "A Decade of Feminist Critiques," 192.

33. R. Bleier, L. Houston, and W. Byne, "Can the Corpus Callosum Predict Gender, Age, Handedness, or Cognitive Differences?" *Trends in Neurosciences* 9 (1986): 391–94; S. Demeter, J. Ringo, and R. W. Doty, "Sexual Dimorphisms in the Human Corpus Callosum," *Abstracts of the Society for Neuroscience* 11 (1985): 868; G. Weber and S. Weis, "Morphometric Analysis of the Human Corpus Callosum Fails to Reveal Sex-Related Differences," *Journal Hirnforschungen* 27 (1986): 237–40; and S. Witelson, "The Brain Connection: The Corpus Callosum Is Larger in Left-Handers," *Science* 229 (1985): 665–68. One of the many complications in making any claims about gender differences appears to lie in the correlation of handedness and age with the size of the corpus callosum. Thus, without controlling for those variations, researchers could find a difference in size between some male and some female corpus callosa, but the difference would not be attributable to the sex difference. Scientists and nonscientists alike should understand that measured differences between two samples or groups may be *real* in the sense that the measurements are not necessarily inaccurate, but the origin of the difference may be from nonrandom sampling (and the nonrandom aspect is not known until other factors are identified—in this case, age and handedness). See Stephen Jay Gould, *The Mismeasure of Man* (New York: Norton, 1981) for examples of actual mismeasurement, some purposeful, some apparently subconscious, along with flawed methodology and conceptualization.

34. Quoted in Bleier, "A Decade of Feminist Critiques," 191. For her critiques, see Bleier, *Science and Gender.*

35. San Francisco *Examiner* (Feb. 22, 1987), as quoted in Bleier, "A Decade of Feminist Critiques," 192–93.

36. *Time* (January 20, 1992).

37. Christien Gorman, "Sizing up the Sexes," *Time* (January 20, 1992): 42.

38. Simon LeVay, "A Difference in Hypothalamic Structure between Heterosexual and Homosexual Men," *Science* 253 (August 30, 1991): 1034; and Marcia Barinaga, "Is Homosexuality Biological?" *Science* 253 (August 30, 1991): 956.

39. LeVay, "A Difference," 1036.

40. Ibid., 1034.

41. Bleier, "A Decade of Feminist Critiques"; and *Science and Gender;* and Fausto-Sterling, *Myths of Gender.* For a more detailed analysis of LeVay's work, see Bonnie Spanier, "Biological Determinism and Homosexuality," *NWSA Journal* 7, no. 1 (Spring 1995): 54–71.

6. From Menstruation to DNA

1. James Darnell, Harvey Lodish, and David Baltimore, *Molecular Cell Biology,* 2d ed. (New York: Scientific American Books, 1990), 11.

2. Emily Martin, *The Woman in the Body: A Cultural Analysis of Reproduction* (Boston: Beacon, 1987), 52–53.

3. Edward H. Clarke, *Sex in Education: or, A Fair Chance for Girls* (Boston: J. R. Osgood, 1873), as cited in Janet Sayers, *Biological Politics: Feminist and Anti-feminist Perspectives* (London: Tavistock, 1982). As Sayers explains (chapter 2: "Sexual Inequality as Reproductive Hazard," 7–27), Clarke's objections to educating women were based on the common but erroneous belief that young women's reproductive development occurred during menstruation. In retrospect, Clarke's concerns, if they were truly in women's interests, could have been countered by letting the young women take time off from studying during their menstrual periods. That Clarke used "biological" arguments only to support his prejudice and worry about women being allowed to attend his Harvard College was obvious from the ludicrous appeals he made in the rest of his writing. In spite of his unsubstantiated claims about the ill effects of studying on a very small number of young women, his message was well received in many quarters, and his book went into seventeen editions through 1905.

4. For example, see the pioneering review of the previous thirty-five years of research, Mary Brown Parlee, "The Premenstrual Syndrome," *Psychological Bulletin* 80 (1973): 454–65. Another early and influential essay was solicited from endocrinologist Estelle Ramey for an early issue of *Ms.* magazine: Estelle P. Ramey, "Men's Cycles (They Have Them Too, You Know)," reprinted in *Beyond Sex-Role Stereotypes,* ed. Alexandra G. Kaplan and Joan P. Bean (Boston: Little, Brown, 1976). See Lynda Birke and Sandy Best, "Changing Minds: Women, Biology, and the Menstrual Cycle," in *Biological Woman—The Convenient Myth,* ed. Ruth Hubbard, Mary Sue Henifin, and Barbara Fried (Cambridge, Mass.: Schenkman, 1982), 161–84; and Anne Fausto-Sterling, *Myths of Gender: Biological Theories about Women and Men* (New York: Basic Books, 1985), chapter 4, "Hormonal Hurricanes, Menstruation, Menopause, and Female Behavior," 90–122.

5. Elliott B. Mason, *Human Physiology* (Menlo Park, Calif.: Benjamin/Cummings, 1983), 525, as quoted in Martin, *Woman in the Body,* 47.

6. Martin, *Woman in the Body,* 47.

7. Ibid., 48.

8. Ibid., 50. A quote from one of the textbooks (Arthur C. Guyton, *Physiology of the Human Body,* 6th ed. [Philadelphia: Saunders, 1984], 498–99) illustrates her point:

> The primary function of the gastric secretions is to begin the digestion of proteins. Unfortunately, though, the wall of the stomach is itself constructed mainly of smooth muscle which itself is mainly protein. Therefore, the surface of the stomach must be exceptionally well protected at all times against its own digestion. This function is performed mainly by mucus that is secreted in great abundance in all parts of the stomach. The entire surface of the stomach is covered by a layer of very small *mucous cells,* which themselves are composed almost entirely of mucus; this mucus prevents gastric secretions from ever touching the deeper layers of the stomach wall.

9. Ibid., 40–41.

10. Ibid., 52–53. The importance of the choice of metaphors to convey gendered or other types of associations and related values is illustrated in a different critique of biology: "The pituitary could be called the 'switchboard' gland (a female gender image) or the 'integrator' gland (a dialectical image)" (Biology and Gender Study Group, "The Importance of Feminist Critique for Contemporary Cell Biology," *Hypatia* 3 [Spring 1988]: 74).

11. James D. Watson et al., *Molecular Biology of the Gene,* 4th ed. (Menlo Park, Calif.: Benjamin/Cummings, 1987), v, vi.

12. Darnell, Lodish, and Baltimore, *Molecular Cell Biology,* 2d ed., 10.

13. *Molecular and Cellular Biology,* 9, no. 2 (February 1989).

14. David Shub, in Hudson Mohawk Association of Colleges and Universities, Program for Faculty, 1990–91.

15. Darnell, Lodish, and Baltimore, *Molecular Cell Biology,* 2d ed., 9; 1st (Scientific American Books, 1986) and 2d eds., 11; 1st and 2nd eds., 11; 1st and 2d eds., 13.

16. Darnell, Lodish, and Baltimore, *Molecular Cell Biology,* 2d ed., 9.

17. Ibid.

18. Ibid., 13.

19. Ibid., 41, 49, 51.

20. Ibid., 53.

21. Ibid., 43.

22. The pattern is created by both the position of the emphasis on the gene as the central focus of molecular biology, and the predominating use of "control" by genes in historical introductions, descriptions of the subject matter, and summaries.

23. Darnell, Lodish, and Baltimore, *Molecular Cell Biology,* 2d ed., 9, 13, 2, 18, 43, 44.

24. For example, ribosomes are referred to as "this complicated protein synthesis machine" in Watson et al., *Gene,* 427.

25. David Baltimore, "The Brain of a Cell," *Science 84* (November 1984): 149–51; quote, 150. I want to thank Scott Gilbert for bringing this article to my attention. See also Donna J. Haraway, "The High Cost of Information in Post–World War II Evolutionary Biology: Ergonomics, Semiotics, and the Sociobiology of Communication Systems," *Philosophical Forum* 13 (1981–82): 244–78, for studies of changes in models of information control after World War II.

26. Baltimore, "The Brain of a Cell," 150.

27. Watson *et al., Gene,* 748–49.

28. Ibid., 704.

29. Nontranscribed controlling sequences of DNA, such as the operator and promoter regions of bacterial operons, are considered a part of the "whole" gene. Darnell, Lodish, and Baltimore, *Molecular Cell Biology,* 2d ed., 342. In addition, the past decade of research on eukaryotes has shown that:

complex transcription units are quite common. Because a transcription unit can produce more than one mRNA and, therefore, encode more than one protein, we must differentiate between the transcription unit and the gene or potential complementation unit. A complex transcription unit may contain two or more polyA sites or two or more splicing variations; this in turn can lead to two or more mRNAs, each of which encodes a separate protein. . . . (343)

30.

The original definition of a cistron implied that it was a DNA region that encodes one polypeptide chain, and many assumed that a cistron would be considered the equivalent of a gene. However, as we shall see, the complicated arrangements of the information in many DNA molecules do not allow every separate polypeptide to be detected by a complementation or recombination test; further, many mutations affect more than one polypeptide. Thus the terms gene and cistron are no longer deemed equivalent.

Molecular Definition of a Gene

The definition of a gene based on classic genetic techniques (i.e., recombination and complementation analysis) cannot account for several observed phenomena—in particular, mutations within a polypeptide-coding region that affect more than one function and mutations located outside a coding region that alter one or more genetic functions. . . . Therefore, a genetically defined cistron is not the "whole" gene; the controlling sequences, which are not transcribed, are also part of the gene.

Ibid., 342.

31. Ibid., 348, 345.

32. Ibid., 43.

33. Charles Mann, "Lynn Margulis: Science's Unruly Earth Mother," *Science* 252 (April 19, 1991): 378; and Lynn Margulis and Dorion Sagan, *Origins of Sex: Three Billion Years of Genetic Recombination* (New Haven: Yale University Press, 1986), 10–15.

34. For a provocative discussion of "informatics of domination," see Donna J. Haraway, "A Manifesto for Cyborgs: Science, Technology, and Socialist Feminism in the 1980s," *Socialist Review* 15 (1985): 65–107.

35. For example:

The more fundamental transformation [rather than the "change from protein to DNA as the carrier of the plan"] in biology from the thirties to the sixties was the working out of the idea of specificity. . . . The structure of DNA then made specificity comprehensible. "Nobody, absolutely nobody, until the day of the Watson-Crick structure, had thought that the specificity might be carried in this exceedingly simple way, by a sequence, by a code," Delbruck said. "This was the greatest surprise for everyone." The rest followed.

Horace Freeland Judson, *The Eighth Day of Creation: Makers of the Revolution in Biology* (New York: Simon and Schuster, 1979), 608, 611–12.

36. Ruth Hubbard, "The Theory and Practice of Genetic Reductionism—From Mendel's Laws to Genetic Engineering," in The Dialectics of Biology Group, *Towards a Liberatory Biology* (New York: Allison & Busby, 1982), 62–78, esp. 68–69. (Revised version, "Genes as Causes," in Ruth Hubbard, *The Politics of Women's Biology* [New Brunswick, N.J.: Rutgers University Press, 1990], 70–86.) A critical (and controversial, even among radical science critics) argument centers on the relationship between a gene and the protein for which it "codes":

The fact that the pattern of inheritance of a qualitative difference between two organisms of the same species can be described by Mendel's Laws permits one to assume that two forms of a gene (alleles) are involved in such a manner that the fact of their difference is registered phenotypically. This does not permit us to conclude that the two forms of the gene are responsible for generating the characters in question. Take normal and sickle-cell haemoglobin, a pair of "characters" whose inheritance follows Mendel's Laws, with normal dominant over sickle-cell. The two molecules are known to differ by a single amino acid. In the simplest case, this difference could be brought about by a change (mutation) in a single base pair in the DNA sequences involved in haemoglobin synthesis. It is correct to say that . . . the difference in the two forms of DNA exerts a decisive effect during the synthesis of the two forms of haemoglobin. It is *not* correct to say that *either* gene *causes* the synthesis of the appropriate form of haemoglobin. Haemoglobin synthesis requires a battery of reactants and energy sources that must come together at the appropriate time and under the appropriate conditions, and that indeed include the appropriate form of DNA. If the DNA is different, a different haemoglobin is formed, or perhaps none if the change in DNA is too great or of the wrong kind. But if a critical substrate concentration is changed, or one of the necessary enzymes, or the temperature, or the pH, or, or, . . . normal haemoglobin also may not be formed. Indeed the cell, or even the organism, may not survive at all. (68)

37. Harold E. Varmus, "Reverse Transcription in Bacteria," *Cell* 56 (March 10, 1989): 721–24.

38. Both Temin and David Baltimore published in 1970 evidence of such activity in RNA tumor viruses, and they subsequently shared the Nobel Prize for that work.

39. Varmus, "Reverse Transcription," 721.

40. Ibid., 723.

41. Evidence that bringing together complementary approaches is valued (at least in *Cell*) is in another mini-review, in which the author lauds the use of several methods and approaches: "a paradigm in complementarity made possible by the succinct, focused application of biochemical, immunological, electron microscopic, and molecular genetics techniques" (Manfred Schliwa, "Head and Tail," *Cell* 56 [March 10, 1989]: 719).

42. Varmus, "Reverse Transcription," 724.

7. Power at a Price

1. James Darnell, Harvey Lodish, and David Baltimore, *Molecular Cell Biology,* 2d ed. (New York: Scientific American Books, 1990), xi–xii.

2. Ruth Hubbard, *The Politics of Women's Biology* (New Brunswick, N.J.: Rutgers University Press, 1990), 116–17.

3. As quoted in Jeffrey L. Fox, "The DNA Double Helix Turns 30," *Science* 222 (October 7, 1983): 30.

4. James Darnell, Harvey Lodish, and David Baltimore, *Molecular Cell Biology,* 1st ed. (New York: Scientific American Books, 1986), viii; (2d ed., xii). The context in which the authors use this key term is as follows:

We hope that the availability of this material in a unified form will stimulate the teaching of molecular cell biology as an integral subject and that such integrated courses will be offered to students as early as possible in their undergraduate education. Only then will students be truly able to grasp the findings of the new biology and its relation to the specialized areas of cell biology, genetics, and biochemistry.

5. Ibid., 2nd ed., xi–xii.

6. This new information includes nonlinearity of genes, post-transcriptional processing of RNA, mobile genetic elements, and no identifiable function for "junk" DNA.

7. This is a good example of the way a technique's "valence" interacts with the context in which the technique is used and given meaning to produce a constraining framework. See n. 16.

8. Darnell, Lodish, and Baltimore, *Molecular Cell Biology,* 2d ed., vii–viii. Here is an example of how a self-fulfilling prophecy contributes to an epistemology or system of knowledge.

9. I stand by that assertion, in spite of the following statement from the final pages of Watson et al., *Molecular Biology of the Gene* (Menlo Park, Calif.: Benjamin/ Cummings, 1987), highlighted by the subheading in bold:

Molecular Biology Is a Subdiscipline of Biology

The evolution of genes is necessarily complex because genes exist only within organisms, and organisms can interact with one another and with the environment in exceedingly subtle ways. Although molecular biology is concerned primarily with molecules, it must always be seen as a subdiscipline of biology. As the next two sections will illustrate, we simply cannot expect to understand the structure and function of genes unless we are prepared to understand the biology of the organisms in which those genes reside. (1157)

The statement is belied by the language and orientation of his textbook and his pronouncements on the importance of the Human Genome Project in understanding "life." Nonetheless, if Watson and his colleagues were to take the above position on the changes in biology that place molecular genetics in an overarching and organizing position, perhaps a counterbalance could be achieved.

10. The last phrase replaced the first two as the title of the chapter. Darnell, Lodish, and Baltimore, *Molecular Cell Biology,* 1st ed., 221; 2d ed., 189. Subsequent quotes from this chapter are from the newer edition, unless otherwise noted.

11. Darnell, Lodish, and Baltimore, *Molecular Cell Biology,* 190.

12.

An avalanche of technical advances in the 1970s drastically changed this perspective. First, enzymes were discovered that cut the DNA from any organism at specific short nucleotide sequences, generating a reproducible set of pieces. The availability of these enzymes, called *restriction endonucleases,* greatly facilitated two important developments: DNA cloning and DNA sequencing.

Ibid., 189–90. A notable change in wording eliminated "The power and the success of the new technology *have given birth to* many hopes for . . ." (1st ed., 222).

13. See chapter 2 and Appendixes B and C.

14. Evelyn Fox Keller, "Physics and the Emergence of Molecular Biology," *Journal of the History of Biology* 23 (1990): 389–409; and "From Secrets of Life to Secrets of Death," in Keller, *Secrets of Life, Secrets of Death: Essays on Language, Gender, and Science* (New York: Routledge, 1992), 39–55, esp. 51. My study provides support for extending to the present Keller's claims about the philosophical origins of molecular biology; see chapter 2.

15. For example, Hilary Rose and Steven Rose, "The Incorporation of Science," in *Ideology of/in the Natural Sciences,* ed. Hilary Rose and Steven Rose (Cambridge: Schenkman, 1979), 16–33; and Sal Restivo, "Modern Science as a Social Problem," *Social Problems* 35 (1988): 206–25.

16. For example, a gun is valenced toward violence, and television is valenced toward viewer passivity. (Corlann Bush, "Women and the Assessment of Technology: to Think, to Be; to Unthink, to Free," in *Machina ex Dea: Feminist Perspectives on Technology,* ed. Joan Rothschild [New York: Pergamon, 1983], 154–55.)

17. Bush specifies four types of contexts that must be taken into account in any feminist analysis of a technique or technology (which she defines as "the organized systems of interactions that utilize tools and involve techniques for the performance of tasks and the accomplishment of objectives"). These are: the design or developmental context, the user context, the environmental context, and the cultural context of "norms, values, myths, aspirations, laws, and interactions of the society of which the tool or technique is a part." (Ibid., esp. 157–58.)

18. Funding for new projects and competing renewals in NIH (National Institutes of Health) study sections has been cut from 6,000 to 4,600 projects/renewals a year, while, among the total grants approved for funding by the study sections, the proportion actually funded has dropped from forty percent to twenty-five percent. Furthermore, all awarded grants had their individual budgets cut by ten to twenty percent. (Bernard D. Davis et al., "The Human Genome and Other Initiatives," *Science* 249 [July 27, 1990]: 342–43.)

19. Bernard D. Davis et al., "The Human Genome." The cosigners, among them Harold Amos, Jonathan R. Beckwith, Alice S. Huang, Ruth Sager, and Priscilla Schaffer, reflect a very broad spectrum of political positions.

20. The concern expressed about continued support for new and ongoing peer-re-

viewed research was also used as a forum for a "deeper" concern of many scientists that stemmed

> from doubts about the scientific justification for the present status of the HGP. Many are not convinced that a crash program for analyzing the structure of genomes will advance either health or the life sciences, for many years to come, as much as studies of specific physiological and biochemical functions and their abnormalities.

Bernard D. Davis et al., "The Human Genome," 342.

Thus, these scientists criticized the proposed project to sequence all of the human genome for its own sake as an inefficient and skewed effort to understand human diseases and physiology.

21. Advertisement, *Science* 243 (January 20, 1989): 328. Another straightforward example of the reorganization of fields is found in another relatively new journal, *Trends in Genetics,* in which development and differentiation are placed as subcategories under genetics, with guiding questions such as: "What controls early development . . . ? How is sex determined? [W]hich genetic processes are involved in evolution? [W]hat defects underlie genetic diseases?" (Alfonso Martinez-Arias, "A Molecular Season for Descriptive Embryology," *Trends in Genetics* 2 [1986]: 146, as quoted in Anne Fausto-Sterling, "Life in the XY Corral," *Women's Studies International Forum* 12 [1989]: 319–33, esp. 319–20, 332.)

22. Advertisement for Vivian Moses and Ronald E. Cape, eds., *Biotechnology, The Science and the Business* (Reading, U.K.: Harwood, 1991) in *Science* 251 (January 11, 1991).

23. Leslie Roberts, "Genome Project Under Way, at Last," *Science* 243 (January 13, 1989): 167–68; quote, 167. James D. Watson was, until recently, the director of the Human Genome Project.

24. See Ruth Hubbard and Elijah Wald, *Exploding the Gene Myth* (Boston: Beacon, 1993), for a close analysis of some of the consequences for medical and health concerns.

25. Darnell, Lodish, and Baltimore, *Molecular Cell Biology,* 1st ed., viii; 2d ed., xii.

26. Ibid.

27. For example, the Darnell, Lodish, and Baltimore textbook was used at the Massachusetts Institute of Technology for a faculty seminar. While science writers probably consult the current science textbooks, Natalie Angier, known increasingly for being enamored of a molecular genetics approach to biology, made specific mention of reading this particular textbook. Natalie Angier, *Natural Obsessions: The Search for the Oncogene* (Boston: Houghton Mifflin, 1988), 173.

28. The view that, politics aside, DNA *is* a "master molecule" is not limited to science novices.

29. The predominant view in *Science,* whether reflected in the editorials, particularly those of Daniel Koshland, Jr., or the assumptions undergirding research and review articles, is that gene cloning and sequencing is the most promising and exciting approach to understanding molecular mechanisms of life. In my analysis in this and other chapters, I point to the exceptions in order to highlight the predominant view and also to illustrate alternative views that are present in the field.

30. Fox, "The DNA Double Helix Turns 30," 29–30.

31.

> "I want to dispel the impression that we know all there is to know," said Sidney Brenner, who is head of the Medical Research Council Laboratory for Molecular Biology in Cambridge, England, in one of the rare moments during the 3-day meeting when the current state of molecular biology was genuinely criticized.

Ibid., 30.

32. Ibid.

33. Ibid.

34. Watson et al., *Gene,* v.

35. Darnell, Lodish, and Baltimore, *Molecular Cell Biology,* 1st ed., 1036.

36. Ibid., 1076. Note that this summary is nearly unchanged in the second edition (998).

37. Ibid., 1036.

38. Ibid., 1075.

39. That this arena is controversial should be no surprise, since so much is at stake in determining what causes cancer. Views on the risks of substances found to be mitogenic and carcinogenic in mice when given in very large doses have recently shifted away from an extremist position that everything causes cancer. Bruce Ames, for one, has changed his position on the kinds of evidence that are useful in ascertaining what promotes cancer, but the complex debate continues. Letters, *Science* 250 (December 21, 1990): 1644–46; and *Science* 251 (February 8, 1991): 606–608.

40. Oncogenes are genes involved in transforming or changing the growth characteristics of cells in tissue culture or in inducing cancer in animals. Antioncogenes are hypothetical genes presumed to be involved in countering the effects of oncogenes. (Darnell, Lodish, and Baltimore, *Molecular Cell Biology,* 996.)

41. Ibid.

42. Ibid.

43. Another example:

Nongenetic mechanisms may play some role in cancer induction. For example, some chemicals, called promoters, can potentiate the activity of electrophilic carcinogens. The best-understood promoters are the phorbol esters, which cause nongenetic changes that often mimic transformation. These substances activate a cellular protein kinase. Long-term treatment with phorbol esters leads to permanent cellular alterations that may or may not be genetic. An apparently clear-cut case of a nongenetic change that causes cancer is the epigenetic alteration leading to a teratocarcinoma. These tumor cells revert to normal when they are implanted into early embryos.

Ibid., 997.

44. Ibid., 995

45. Ibid.

46. Ibid., 996

47. National Women's Health Network communication, June 1993; and Ann Gibbons, "Women's Health Issues Take Center Stage at the IOM," *Science* 258 (October 30, 1992): 733.

48. Darnell, Lodish, and Baltimore, *Molecular Cell Biology,* 994–96.

49. Ibid., 998.

50. Ibid., 956.

51. Cast in leftist language, the central assertion is clear and, whether or not you agree with the political stance, quite accurately highlights the power of the framework or paradigm in shaping answers:

But whether the cause of tuberculosis is said to be a bacillus or the capitalist exploitation of workers, whether the death rate from cancer is best reduced by studying oncogenes or by seizing control of factories—these questions can be decided objectively only within the framework of certain sociopolitical assumptions.

Richard Levins and Richard Lewontin, *The Dialectical Biologist* (Cambridge: Harvard University Press, 1985), 5.

52. See chapter 2; and Scott Gilbert, "Intellectual Traditions in the Life Sciences: Molecular Biology and Biochemistry," *Perspectives in Biology and Medicine* 26 (1982): 151–52.

53. Hubbard, *Politics,* 116–17.

54. Ibid.

55. Rita Levi-Montalcini, "The Nerve Growth Factor 35 Years Later," *Science* 237 (September 4, 1987): 1154–62; quote, 1157. Dr. Levi-Montalcini is at the Institute of Cell Biology, NRC, Rome, Italy.

56. Ibid., 1159.

57. Ibid., 1160.

58. Ibid., 1158, 1161.

59. Ibid., 1158.

60. Darnell, Lodish, and Baltimore, *Molecular Cell Biology,* viii.

61. Ibid., 1st ed., 992. See chapter 2 for a discussion of slime molds as models for studying differentiation.

62. Ibid.

63. Ibid., 2d ed., 751–52.

64. This is the issue addressed by Keller in her critique of the pacemaker model (of centralized control and inherent difference) as the one preferred to an equally plausible model of steady-state dynamics in which difference is generated by random environmental changes and responses of cells; see my chapter 2. Evelyn Fox Keller, "The Force of the Pacemaker Concept in Theories of Aggregation in Cellular Slime Mold," in *Reflections on Gender and Science* (New Haven: Yale University Press, 1985), 150–57.

65. Ruth Herschberger, *Adam's Rib* (New York: Harper and Row [1948] 1970).

66. Susan Griffin, *Woman and Nature: The Roaring inside Her* (New York: Harper and Row, 1978).

67. As, I believe, Griffin's work is misunderstood in Evelyn Fox Keller, "Women, Science, and Popular Mythology," in *Machina ex Dea,* 130–46.

68. See Donna J. Haraway, "Manifesto for Cyborgs: Science, Technology, and Socialist Feminism," *Socialist Review* 80 (1985): 65–107, on the significance of irony for feminist social change.

69. Starhawk, *The Spiral Dance: A Rebirth of the Ancient Religion of the Great Goddess* (San Francisco: Harper and Row, 1979).

70. See Watson et al. *Gene,* 1159–60; Rebecca L. Cann, Mark Stoneking, and Allan C. Wilson, "Mitochondrial DNA and Human Evolution," *Nature* 325 (1987): 31–36; Rebecca L. Cann, "In Search of Eve," *The Sciences* (September/October 1987): 30–37; Henry Gee, "Statistical Cloud over African Eden," *Nature* 355 (1992): 583; and Alan G. Thorne and Milford H. Wolpoff, "The Multiregional Evolution of Humans," *Scientific American* 266 (April 1992): 76–83.

71. Darnell, Lodish, and Baltimore, *Molecular Cell Biology,* 2d ed., 1053.

72. For example, see Donna Haraway's discussion of Octavia Butler's futuristic novels (Donna J. Haraway, *Primate Visions* [New York: Routledge, 1989], 376–82); also Marge Piercy, *Woman on the Edge of Time* (New York: Fawcett Crest, 1976); and Ursula K. Le Guin, *The Left Hand of Darkness* (New York: Walker, 1969).

8. Molecular Biology

1. James D. Watson and John Tooze, *The DNA Story: A Documentary History of Gene Cloning* (San Francisco: W. H. Freeman, 1981), 584.

2. Sheldon Glashow, Nobel laureate, theoretical physicist, and self-described "unreformed reductionist and an unredeemed positivist," in his lecture at the 25th Annual Nobel Conference, Gustavus Adolphus College, "The End of Science," quoted in "Does

Ideology Stop at the Laboratory Door? A Debate on Science and the Real World," *New York Times* (October 22, 1989).

3.

> We affirm that there are eternal, objective, extrahistorical, socially neutral, external and universal truths and that the assemblage of these truths is what we call physical science.

Ibid.

4. These generalizations are based on the past decade of experience with reactions to feminist and other critical science studies, including reviews by scientists of grant proposals and conversations with science faculty about teaching science-and-society issues.

5. Helen E. Longino, *Science as Social Knowledge: Values and Objectivity in Scientific Inquiry* (Princeton, N.J.: Princeton University Press, 1990), 7.

6. Ibid., 5 (emphasis added).

7. In chapters 4, 5, 6, and 7, I address other science-and-society issues, such as scientific support for biological determinism, with connections to racism and sexism.

8. James Darnell, Harvey Lodish, and David Baltimore, *Molecular Cell Biology,* 2d ed. (New York: Scientific American Books, 1990), x.

9. Ibid., 1.

10. "The power and success of the new technology have raised high hopes that the practical use of our ever-increasing biological knowledge will bring many benefits to mankind." (Ibid., 190; compared to 1st ed., 222). Subtle changes from the first edition to the second are worth noting. They include: adding "how proteins or domains of proteins function"; changing "have given birth to" to "raised high hopes"; changing "human beings" to "mankind"—this from gender-neutral to pseudogeneric male.

11. I reproduce these sections to illustrate the use of the quoted terms in context:

> Since the late 1970s, however, new techniques have revolutionized the study of DNA and the genes that it embodies. The DNA from any organism can be cut into reproducible pieces with site-specific endonucleases called restriction enzymes; the pieces can then be linked to bacterial vectors (DNA viruses or plasmids capable of independent growth) and introduced into bacterial hosts. In this way, almost any DNA segment from any organism can be isolated and produced in any amount simply by culturing the bacteria. These procedures are collectively referred to as gene cloning or recombinant DNA technology. Rapid DNA sequencing techniques were worked out. Suddenly, biologists were ushered into a new era: instead of just counting chromosomes or identifying gene mutations and mapping genes imprecisely by means of breeding experiments, they could now obtain the exact DNA sequence of individual genes and thus their exact location and coding capacity.
>
> Not only has the isolation of particular DNA segments and their sequencing become commonplace, but molecular biologists can now introduce specific mutations at will; altered genes can be reintroduced into cells or even into the germline of organisms, so that the function of individual genes can be studied in particular cells. These fantastic strides forward, combined with the ever-improving methods for examining cells and their components, both biochemically and structurally, have aroused great optimism about the success of future research. We stand on the brink of the formerly unthinkable achievement of knowing the DNA sequence of the entire human genome, perhaps by the end of this century. Surely few problems in molecular cell biology, even finding the cause for cancer or determining the molecular basis of differentiation and development, will remain unsolved much longer.

Ibid., 1; 14. Again, changes from the first edition are interesting. In the last sentence quoted in the text, "finding a cure for cancer" was changed to the more modest "finding the cause for cancer."

[Are certain cells predetermined in embryological development or does the environment of the cell influence its fate?] Only the identification of the key genes that define the activities of two different cell states and an analysis of what controls the earliest expression of these genes will shed light on such problems.

Ibid., 1st ed., 992. With the reorganization of the second edition to incorporate developmental biology throughout, rather than devoting a separate chapter to it, this statement seems to have been deleted.

12. In several instances, changes in wording from the first edition tone down the limitless enthusiasm, presenting a more evenhanded description of developments in the field. For example, in the first edition:

Industrial microbiologists can use recombinant DNA techniques to engineer bacteria and other easily cultured organisms to make proteins that are used in medicine, in agriculture, and in research of all kinds. Progress in this area has been astonishingly rapid. (255)

The second edition has dropped the last sentence and added a list of the proteins produced with those methods (2nd ed., 218–19). Even with these changes, however, no reference is made to possible detrimental consequences, biohazards, ethical issues, etc., from applying recombinant DNA technology.

13. Everett Mendelsohn, " 'Frankenstein at Harvard': The Public Politics of Recombinant DNA Research," manuscript, 1980; and Charles Weiner, "Historical Perspectives on the Recombinant DNA Controversy," in Joan Morgan and W. J. Whelan, eds., *Recombinant DNA and Genetic Experimentation* (New York: Pergamon, for COGENE, 1979). Much has been written on this subject. A revealing publication with regard to scientists' representations of themselves in scientific controversies is the AAAS publication in the Issues in Science and Technology Series, *The Gene-Splicing Wars: Reflections on the Recombinant DNA Controversy,* ed. Raymond A. Zilinskas and Burke K. Zimmerman (New York: Macmillan, 1986).

14. A significant point of contention was the effort, made by David Baltimore among others, to segregate

the scientific issues from ethical and moral issues. The meeting was not held to take up "peripheral" issues such as genetic engineering and its potential for chemical and biological warfare. Instead, the aim was to develop a strategy for research which would maximize benefits and minimize hazards.

Mendelsohn, " 'Frankenstein at Harvard,' " 5

15. A special issue of *Daedalus, Journal of the American Academy of Arts and Sciences* (vol. 107, Spring 1978) was devoted to "Limits of Scientific Inquiry." In it, the pros and cons of placing limits on scientific research were debated from several perspectives. David Baltimore was among those strongly against limitations derived from particular ideologies:

[S]hould limits be placed on biological research because of the danger that new knowledge can present to the established or desired order of our society? . . . I believe that there are two simple, and almost universally applicable, answers. First, the criteria determining what areas to restrain inevitably express certain sociopolitical attitudes that reflect a dominant ideology. Such criteria cannot be allowed to guide scientific choices. Second, attempts to restrain directions of scientific inquiry are more likely to be generally disruptive of science than to provide the desired specific restraints. These answers to the question of whether limits should be imposed can be stated in two arguments. One is that science should not be the servant of ideology, because ideology assumes answers, but science asks questions. The other is that attempts to make science serve ideology will merely make science impotent without assuring that only desired questions are investigated. I am stating simply that we should not control the direction of science and, moreover, that we cannot do so with any precision.

Baltimore, "Limiting Science: A Biologist's Perspective," 41
The Nazis' use of eugenics for racial purity and the Soviet Union's dedication in the 1920s to Lamarckian heredity of acquired characteristics were common examples given of the distortion of science by political ideology. Lacking are discussions of the ways that the "dominant ideology" shapes "good" science.

16. James D. Watson et al., *Molecular Biology of the Gene,* 4th ed. (Menlo Park, Calif.: Benjamin/Cummings, 1987), v, vi. There are, not surprisingly, other judgments embedded in this text. For example, the pressures of time on research progress are exaggerated beyond the currently high pressures created by the scientific establishment. This belief in and reinforcement of a highly time-competitive system selectively works against certain groups: anyone desiring to research on less than a full-time basis (mostly women with childbearing and child-rearing responsibilities); anyone who cannot afford to do undergraduate or graduate work full-time (lower socioeconomic groups, which are disproportionately composed of minorities and women, but which also include white working-class and poor men); scientists with positions at liberal arts or community colleges, rather than research universities, etc. A humorous reading of the problem that molecular biology has outgrown the possibility of being taught from "a handy volume that would be pleasant to carry across a campus" points to the size and weight of massive textbooks in the sciences, suggesting that only football players are capable of gaining access to molecular biology and related sciences. Pleasantness also contrasts with the seriousness of the business of becoming a molecular biologist.

17. Ibid., v.

18. Zilinskas and Zimmerman, *Gene-Splicing Wars,* 224.

19. Darnell, Lodish, and Baltimore, *Molecular Cell Biology,* 2d ed., x.

20. For example, Alice Kimball Smith, *A Peril and a Hope: The Scientists' Movement in America 1945–47* (Chicago: University of Chicago Press, 1965).

21. Susan Wright, "Molecular Biology or Molecular Politics? The Production of Scientific Consensus on the Hazards of Recombinant DNA Technology," *Social Studies of Science* 16 (1986): 593–620, esp. 604.

The assumptions that Wright claims narrowed the discussion to a limited number of less problematic hazards include:

(a) that only one weakened strain of bacteria, *E. coli* K12, would be used for all recombinant DNA research;

(b) that only hazards outside the laboratory were to be considered, brushing aside the issue of potential hazards to workers inside labs; and

(c) that recombinant DNA research would occur only in technologically advanced countries with conditions (public health and sewage treatment) that would minimize epidemics. (602–604)

22. Ibid., 604. The reference to equally dangerous work is to microbiological experiments with pathogenic organisms. For example, in the 1970s, a photographer in England died from the escape of smallpox virus from a research lab to another floor of a research building.

23. Ibid., 606.

24. Watson and Tooze, *DNA Story,* 561, 584.

25. Ibid.

26. Ibid.

27. Sheldon Krimsky, *Biotechnics and Society: The Rise of Industrial Genetics* (New York: Praeger, 1991).

28. Watson and Tooze, *DNA Story,* 584.

29. Ibid.

30. For example, Sheldon Krimsky, *Genetic Alchemy: The Social History of the Recombinant DNA Controversy* (Cambridge: MIT Press, 1982); and Zilinskas and Zimmerman, *Gene-Splicing Wars.* While the latter is a collection of essays reflecting a range of perspectives, the preface clearly trivializes one side by featuring the story of Chicken Little, with the conclusion: "But the sky never fell. . . . Foxy Loxy, whose livelihood depends on mass hysteria, is occasionally seen filing suit against the federal government or trying to get people to believe that the genetic heritage of a cow is an inviolable and sacred right." (ix)

31. In her report entitled, "Telling the Truth," Lynne V. Cheney, President Bush's head of the National Endowment for the Humanities, charged: "The aim of education, as many of our campuses now see it, is no longer truth, but political transformation—of students and society." "Liberal scholars are using college classrooms to advance their political agendas and indoctrinate their students . . . Mrs. Cheney especially blamed feminists for pushing their agendas on students." (Stephen Burd, "Humanities Chief Assails Politicization of Classrooms," *Chronicle of Higher Education* [September 30, 1992]: A21–22).

32. Let me reiterate that my statements are generalizations about the dominant modes of science education and are not blanket criticisms of science educators. There are certainly many individual science teachers who teach from a different set of beliefs and concerns, and they should be acknowledged as models. But few departments or institutions can be found that provide a consistent alternative to dominant socialization in the natural sciences. Important exceptions, fueled by concerns about the discouragement of women and minority students from science, include the science curriculum at Hampshire College, which has offered such first-year seminars as The Biology of Women, Human Movement in Physiology, and Ethical Issues in Biomedical Research, a seminar for science majors about women in science, as well as a pedagogy that encourages collaboration and collective learning about scientific topics of particular concern to societal problems. (Mary Jo Strauss, "Feminist Education in Science, Mathematics, and Technology," *Women's Studies Quarterly* 11, no. 3 [Fall 1983]: 23–25.)

33. Watson and Tooze, *DNA Story,* viii.

34. Ibid., vii.

35. In striking, but not surprising, contrast, Donna J. Haraway opens *Primate Visions* by deconstructing the binary opposition of fact and fiction, instead casting science as just one mode of story-telling. Not intending to trivialize scientific knowledge, Haraway also acknowledges insights from certain forms of science fiction (such as Octavia Butler's writings) and other stories less bound by the cultures of Western science and patriarchy. (Donna J. Haraway, *Primate Visions* [New York: Routledge, 1989], esp. 3–5.)

36. Benjamin Lewin, editorial, "Travels on the Fraud Circuit," *Cell* 57 (May 19, 1989): 513–14. The event that prompted this was a meeting that brought together scientists and politicians' staffers to address scientific fraud.

37. Ibid., 513.

38. Ibid.

39. Arthur Caplan, "The World of Science Pays a High Price for Honesty," *Times-Union* (Albany) (April 7, 1991): D-2 (Caplan is the director of the Center for Biomedi-

cal Ethics at the University of Minnesota); David P. Hamilton, "Did Imanishi-Kari Get a Fair 'Trial'?" *Science* 252 (June 21, 1991): 1607 (that case is a quagmire in many ways, but it may have encouraged more scientists to consider some of the factors contributing to the problem, such as ever larger labs); Marcia Barinaga, "Labstyles of the Famous and Well Funded," *Science* 252 (June 28, 1991): 1776–78.

40. Lewin, "Travels," 514.

41. I suggest that the level of concern in Lewin's editorial stems from a realization that scientific fraud is far more widespread than most scientists would like to believe or, certainly, to acknowledge. Anyone familiar with current practices in molecular biology, as one example of a well-funded area of science, knows that the very large size of laboratories, with many more graduate students, postdoctoral fellows, and technicians than any one human being can keep after; the pressures to publish; and the increased competition for grant support create fertile ground for fraud—that may not even be considered fraud in the view of some scientists. Some scientists are so certain of the answer they will find that doctoring the data does not seem like outright fraud. They may believe that with enough experiments they would get the answer they are publishing, but they have neither the time nor the resources to repeat the experiments. I suggest that scientists must do more than set up mechanisms for more satisfactory handling of allegations. We must begin to address the issues hidden within the fraud cases: too large, too much at stake, not enough resources for different approaches to a problem, etc.

42. Criticism has been leveled against Lewin in particular, but also against the editors-in-chief of *Nature* and *Science,* that they use their position to rush to publication certain papers in molecular biology involved in priority claims. (Leslie Roberts, "The Rush to Publish," *Science* 251 [January 18, 1991]: 260–63.)

43. Daniel E. Koshland, Jr., editorial, *Science* (January 6, 1989): 9.

44. Thomas J. Bouchard, Jr., et al., "Sources of Human Psychological Differences: The Minnesota Study of Twins Reared Apart," *Science* 250 (Oct. 12, 1990): 223–28. This issue also features "The Human Map" and includes a wall chart of the Human Genome Project's progress toward mapping the twenty-three pairs of chromosomes in humans. Which humans? The cover of the issue portrays a human form that is more typical of males in this culture: muscular neck, shoulders, and arms (it could indeed be a woman, as there are many women in this and other cultures with well-developed musculature in those places, while there are many men with much less muscle development; however, to the acculturated eye, the form is that of a male). The collage also includes a photo of a man superimposed over a genetic pedigree chart, as well as an outline of a mouse. Women are nowhere in evidence.

45. R. C. Lewontin, Steven Rose, and Leon J. Kamin, *Not In Our Genes: Biology, Ideology, and Human Nature* (New York: Pantheon, 1984), 100–106, 118–19, 212. See also L. S. Hearnshaw, *Cyril Burt, Psychologist* (London: Hodder and Stoughton, 1979).

46. Bouchard et al., "Minnesota Study," 228.

47. "This Week in Science," *Science* 250 (October 12, 1990): 187; and Daniel E. Koshland, Jr., editorial, "The Rational Approach to the Irrational," *Science* 250 (Oct. 12, 1990): 189.

48. Koshland, "Rational Approach," 189. Koshland had been criticized earlier that year (Letters: Maurice Fox, Boris Magasanik, Ethan Signer, Frank Solomon, Martin Gellert, and James Haber, *Science* 247 [January 19, 1990]: 270) for his "enthusiasm for the use of human genome mapping in social policy" and for vilifying opponents of the project. His response was that he was misunderstood.

49. Paul R. Billings, "Promotion of the Human Genome Project," *Science* 250 (November 23, 1990): 1071. Koshland again replies that he has been misunderstood. But he reiterates his major assertions about the usefulness of genetics in curing mental

illness: "The knowledge will be just as useful and no more so than that applied to other inherited illnesses such as cystic fibrosis."

50. Compare the near absence of discussions of potential problems in an issue devoted to "The New Harvest: Genetically Engineered Species," *Science* 244 (June 16, 1989): "Few discussions of gene therapy at scientific meetings and in publications still argue its need or potential place in medicine or its ethical acceptability. . . ." (Theodore Friedmann, "Progress toward Human Gene Therapy," 1275) to H. Patricia Hynes's contextual analysis of such technologies ("Biotechnology in Agriculture: An Analysis of Selected Technologies and Policy in the United States," *Reproductive and Genetic Engineering* 2 [1989]: 39–49).

51. One of many examples:

Now on the threshold of [biology's] first large organized project, it is certain that biologists in particular and humanity in general will obtain remarkable benefits over the next 15 years, with relatively insignificant risk.

Charles C. Cantor, "Orchestrating the Human Genome Project," *Science* 248 (April 6, 1990): 51.

52. The letter is written in the calm and reasoned tone of scientists and academics, but its position is quite clear:

We biologists are committed to using the new genetic technology for diagnosing, curing, and preventing disease, not causing it, as well as for such purposes as the improvement of agricultural crops, reversal of genetic disease, provision of rare biochemicals and the unravelling of biological mechanisms. We abhor the use of biological agents as offensive weapons by any nation, in accord with the many nations who signed the 1972 International Convention banning the use or stockpiling of biological weapons.

   Although we recognize DOD's responsibility to provide defense against possible biological attack, we find their program to be flawed, hazardous, and likely to break the constraints of the 1972 Convention. . . . In the second place, an infinite variety of potentially lethal agents already exist or could be produced by genetic engineering; engineered organisms raise the specter of epidemics that can be neither diagnosed nor treated.

Naomi C. Franklin et al., "Petition on Dugway Facility," *Science* 243 (January 6, 1989): 11 (letters section).

53. Treatment of biological determinist reports has actually *narrowed* to exclude alternate perspectives. About ten years ago, *Science* published an article claiming that observed differences between girls and boys in performance on math SATs were due primarily to inherent differences between the sexes, because Benbow and Stanley had controlled for math courses taken and thus eliminated that major criticism of previous studies. In the News section of the same issue, however, those researchers' conclusions were labeled controversial by citing other investigators with very different assumptions and data about math ability and the sexes. What Benbow and Stanley attributed to a theoretical "endogenous" variable could more verifiably be ascribed to the influence of cultural attitudes about mathematics and gender—beliefs shown to be held by parents, teachers, counselors, and the students themselves. Thus, in 1980 *Science* gave some space to the "factors" ignored or trivialized by Benbow and Stanley. Three years later, *Science* published another study of theirs, but without additional comment, and the title included the term, "More Facts." (See my chapter 3.)

54. David T. Suzuki et al., *An Introduction to Genetic Analysis,* 4th ed. (New York: W. H. Freeman, 1989), 1. The full list of key concepts reads:

Genetics has unified the biological sciences.
Genetics may be defined as the study of genes through their variation.
Gene variation contributes to variation in nature.
The characteristics of an organism are determined by the interaction of its unique
  set of genes with its unique environment.
Genetics is of direct relevance to human affairs.

55. Ibid., 12.
56. Ibid.
57. Ibid., 13. Those problems as articulated by the authors:

The use of the new high-yield varieties produced a wide range of social and economic problems in the impoverished countries where they were most needed.
Furthermore, the spread of monoculture (the extensive reliance on a single plant
variety) left vast areas at the mercy of some newly introduced or evolved form of
pathogen. . . .

Thus, students are taught that creating new genetic varieties of food crops is not sufficient to eradicate food shortages.
58. The authors do not limit themselves to one side of the debate; they even cite
some of the problematic books that present a highly critiqued view of social behavior
and genetics:

The nature of inborn constraints on thought and personality and the implications
for present sociological problems have fascinated many geneticists and other scientists. Such books as *African Genesis, The Territorial Imperative, On Aggression,* and
other popular titles on sociobiology have stimulated widespread public interest. A
bitter debate has raged about the differences in intelligence among various racial
and social groups. . . . There is very serious talk, even among some geneticists,
about the ability of the human race to take control of its own evolution. Others
are frightened by the possibilities for disastrous error or unacceptable sociological
consequences.

Ibid., 13–14.
59. Ibid., 15. Specific books presenting the argument for potential dangers from
gene-splicing applications are named.
60. Ibid.
61. Ibid., 14, figure 1-13.
62. Ibid., 6, 7, 10.
63. Scott F. Gilbert, *Developmental Biology,* 2d ed. (Sunderland, Mass.: Sinauer Associates, 1988).
64. Ibid., xv.
65. Richard Davenport, *An Outline of Animal Development* (Reading, Mass.: Addison-
Wesley, 1979), 4. See my chapter 7 for Ruth Hubbard and Niels Bohr on this view.
66. Ibid., 1–2.

The discussion presented here, while not necessarily antithetical to the preceding
goals, is based on the conviction that the point of view from which such goals
and approaches are derived does not provide an adequate basis for an understanding of ontogeny [development]. In fact, it can obscure rather than illuminate.

Davenport raises questions about what he sees as an extreme position on reductionism
in science today and the consequences for the quality of life and of scientific epistemology.

Too often we forget that modern science evolved out of the rationalism of the Middle Ages, which was characterized by a faith based on reason, and has become, instead, a reason that is based on faith—faith in the ultimate orderliness of Nature. By revolting against the rationalism of medieval thought, we have insisted on reducing Nature to what Whitehead has termed "irreducible and stubborn facts." However, this reduction and its assumptions have proved to be an overreaction to past folly and have produced an unwarranted devaluation of our experience that must ultimately fail as an impartial representation of Nature.

67. Ibid., esp., 1–5, 351–403.

68. Mark G. Simkin, *Discovering Computers* (Dubuque, Iowa: William C. Brown, 1990), 5.

Conclusion

1. National Academy of Sciences Committee on the Conduct of Science, *On Being a Scientist* (Washington, D.C.: National Academy Press, 1989), 8.

2, Johnnella Butler et al., "Women's Studies Task Force Report," in *Reports from the Fields—Project on Liberal Learning, Study-in-Depth, and the Arts and Sciences Major,* vol. 2 (Washington, D.C., 1990), 211.

3. The Science, Technology, and Gender Symposium was cosponsored by Iowa State University and the National Women's Studies Association as part of the NWSA National Conference, June 16–19, 1994.

4. Title and subheading of news article, Constance Holden, "Wanted: 675,000 Future Scientists and Engineers," *Science* 244 (June 30, 1989): 1536.

5. Ruth Bleier's experience with her critiques of biological determinist claims about sex differences in the brain is one of many examples (see chapter 5). Simon LeVay, the first scientist to come out as gay in *Science,* in connection with his claim that homosexual men have a different brain compared to heterosexual men and women (see chapter 5), is another example of a scientist whose lack of objectivity might affect his science (Eliot Marshall, "When Does Intellectual Passion Become Conflict of Interest?" *Science* 257 [July 31, 1992]: 620). No similar charge was raised against purportedly heterosexual researchers—or male scientists—in the field of sex differences work. Another example is hinted at in Charles Mann, "Lynn Margulis: Science's Unruly Earth Mother," *Science* 252 (April 19, 1991): 378.

6. See my critique in chapter 8 of Thomas Bouchard's twin studies and *Science* editor Daniel Koshland, Jr.'s comments. Also, a subtly racist commentary, Robert Pool, "Who Will Do Science in the 1990s?" *Science* 248 (April 27, 1990): 433–35.

7. "Women in Science: 1st Annual Survey," *Science* 255 (March 13, 1992): 1365–89; and "Minorities in Science: The Pipeline Problem," *Science* 258 (November 13, 1992): 1175–1237.

8. "Women in Science '93: Gender and Culture," *Science* 260 (April 16, 1993): 383–432.

9. Marcia Barinaga, "Feminists Find Gender Everywhere in Science," *Science* 260 (April 16, 1993): 392–93.

10. Case studies in biology support the idea that scientists trying to advance heterodox claims are pressured to argue in terms consistent with accepted beliefs and to avoid directly challenging large areas of accepted theories. (Greg Myers, *Writing Biology: Texts in the Social Construction of Scientific Knowledge* [Madison: University of Wisconsin Press, 1990].)

11. Evelynn Hammonds's analysis of women scientists' attitudes about feminist views of gender and science problems shows that, like their male colleagues, women scientists themselves tend to stop short at the suggestion of a causal relationship

among various forms of gender ideology in the content, process, and organization of science. (Helen E. Longino and Evelynn Hammonds, "Conflicts and Tensions in the Feminist Study of Gender and Science," in *Conflicts in Feminism,* ed. Marianne Hirsch and Evelyn Fox Keller [New York: Routledge, 1990], 164–83, esp. 176–81.)

12. Ruth Bleier, "A Decade of Feminist Critiques in the Natural Sciences," *Signs* 14 (Autumn 1988): 193. See the introduction and dedication of that paper: Judith Walzer Leavitt and Linda Gordon, "A Decade of Feminist Critiques in the Natural Sciences: An Address by Ruth Bleier," *Signs* 14 (Autumn 1988): 182–85. See also Ruth Bleier, "The Cultural Price of Social Exclusion: Gender and Science," *NWSA Journal* 1 (1988): 7–19.

13. Anne Fausto-Sterling, "Building Two-Way Streets: The Case of Feminism and Science," *NWSA Journal* 4, no. 3 (Fall 1992): 336–49; Ruth Hubbard et al., "Comments on Anne Fausto-Sterling's 'Building Two-Way Streets'," *NWSA Journal* 5, no. 1 (Spring 1993): 45–48; Lee Swedberg, "Fallible or Lovable: Response to Anne Fausto-Sterling's 'Building Two-Way Streets'," *NWSA Journal* 5, no. 3 (Fall 1993): 389–91.

14. From Maria Mitchell, "Astronomer," in *Growing Up Female in America: Ten Lives,* ed. Eve Merriam (New York: Dell, 1973), 96. Reprint of Doubleday, 1971.

15. Ruth Hubbard, *The Politics of Women's Biology* (New Brunswick, N.J.: Rutgers University Press, 1990), 32.

INDEX

BONNIE SPANIER is Associate Professor in the Women's Studies Department at the State University of New York, University at Albany. Her doctorate is from Harvard University, Department of Microbiology and Molecular Genetics. She coauthored *Toward a Balanced Curriculum* and has published articles on the molecular biology of Newcastle disease virus and on feminist critiques of the natural sciences. She is a consultant for many colleges and curriculum transformation projects on gender, race, class, and sexuality issues in the sciences.